George Leslie, William Abbott Herdman

The Invertebrate Fauna of the Firth of Forth

George Leslie, William Abbott Herdman

The Invertebrate Fauna of the Firth of Forth

ISBN/EAN: 9783337272975

Printed in Europe, USA, Canada, Australia, Japan

Cover: Foto ©berggeist007 / pixelio.de

More available books at **www.hansebooks.com**

THE

INVERTEBRATE FAUNA

OF THE

FIRTH OF FORTH.

BY

GEORGE LESLIE
Demonstrator of Zoology, University of Edinburgh;

AND

W. A. HERDMAN, D.Sc., F.L.S.

EDINBURGH:
Printed by M'Farlane & Erskine.
1881.

PREFACE.

During the Session 1880-81 we communicated to the Royal Physical Society of Edinburgh a series of papers, in which an attempt was made to catalogue the Invertebrata of the Firth of Forth. While these were being printed for the Society, it was suggested that the lists should appear in a separate form, and the Council have sanctioned their publication before the appearance of the sixth volume of the *Proceedings*, in which they will be printed. We have to thank the Society for this permission, and we are especially indebted to its accomplished secretary, Mr Robert Gray, F.R.S.E., who has, both in his official capacity and as our friend, in many ways forwarded our wishes.

When we began this work, it was intended that the lists should comprise only the results of our own collecting and dredging, but we soon found that if they were to be of general use to the student they must include the results of previous work, of which we have accordingly availed ourselves to such an extent that the lists as they now stand are, to a very large extent, a compilation from the literature of the subject. Little apology is necessary for this assumption of the labour of others, since, with the exception of the lists of Drs M'Bain and Howden, the records of the Forth Fauna are fragmentary, and are only to be found in works often inaccessible to the general student.

We have the pleasant task of acknowledging the valuable assistance of many marine zoologists, whose names are mentioned in connection with their contributions in various parts of this work. Any further help with which we may be favoured in our endeavours to complete the record of the Fauna of the Forth will be gratefully received.

We have not followed the zoological order, the various groups being arranged in the order in which it was found most convenient to undertake them. If they prove useful, we intend to re-issue the lists, when this and other imperfections of the present work will be corrected.

THE INVERTEBRATE FAUNA
OF THE
FIRTH OF FORTH.

INTRODUCTION.

THE following is intended to be the first of a series of papers, in which a list of the Invertebrata found in the Firth of Forth shall be given. We have been led to undertake this work on the suggestion of Sir Wyville Thomson and others, who think it is desirable that a fairly complete record of the marine fauna of the estuary should be easily attainable by those studying the zoology of the eastern coast. The special necessity for a list of the fauna of the Forth will be admitted, when it is considered that the estuary has always been the favourite and most accessible collecting ground with the numerous students of natural history attending the University of Edinburgh, and special facilities for its completion exist, as many able zoologists have during the last two centuries published the results of their work on the same subject. In recent years, however, since zoological classification has assumed its present form, nothing aiming to be a complete revision of the Invertebrata of the Forth has been produced, and the varying nomenclature of species renders the older partial lists of comparatively little use to the student of the present day. We are fully conscious that we will not be able to catalogue the entire invertebrate fauna in these papers, and that even some forms which have been found and chronicled may have escaped our notice; but this is the less to be regretted, as we hope in subsequent papers to add to our lists, or otherwise amend them.

A good description of the physical geography of the estuary may be found in the introduction to Parnell's "Fishes of the Firth of Forth"* and elsewhere, so that we shall here note only some of its more salient features.

* Trans. Wern. Soc., vol. vii., p. 162.

The length of the estuary, measured from its junction with the sea to the vale of Stirling, where it terminates in the river, is about 56 miles. The tide, however, flows to Craigforth, 25 miles above the proper estuary. Where it joins the sea, its breadth, from St Abb's Head on the south side to Fifeness on the north, is nearly 40 miles. As we ascend it rapidly contracts, so that at Elie, 10 miles above Fifeness, it is only 7 miles broad. Above this it again expands into an extensive basin, which at Musselburgh has a breadth of about 20 miles, and then gradually narrows, until at Queensferry it is only 2 miles across. Above Queensferry it again expands, for a distance of about 14 miles, into a basin having an average breadth of 4 miles, and terminates in the river.

The most important islands of the Firth are the May, situated a little above its junction with the sea; the Bass, nearly opposite North Berwick; Inchkeith, in the centre of the greater basin; and nearer its upper contraction the islands of Inchcolm, Inch Mickery, and Cramond Island. Inch Garvie lies opposite Queensferry.

The maximum depth of the estuary at its mouth is about 35 fathoms; at the mouth of the greater basin 28 fathoms. From the May to Inchkeith, the greatest depth in the middle of the channel is from 16 to 18 fathoms. Above Inchkeith the Middle Bank stretches, separating the north channel, with a depth of 16 to 25 fathoms, from the south channel or Leith Roads, varying from 3 to 16 fathoms. At the upper contraction of the greater basin, between Inch Garvie and North Queensferry, the depth increases to 37 fathoms, and above this it gradually becomes shallower. The greater part of our dredging has been done in the south channel, and between Inchkeith and the May.

In 1710, Sir Robert Sibbald, a learned Edinburgh physician, published a list of the exsanguous (invertebrate) animals which he supposed were common to the Firths of Forth and Tay.* He divides them into four classes: (1.) the *Molles*, among which are to be found Cephalopods, Asterids, and Medusæ; (2.) the *Crustrate*, comprehending Crustacea and

---

* "History of Fife and Kinross," p. 53. Edin., 1710.

Echinids; (3.) the *Testacea*, being Gastropods, Lamellibranchs, and Cirripedes; and (4.) the *Sea-insects*, among which we recognise *Aphrodite aculeata*. Sir Robert enumerates sixty-eight species of invertebrates. Of these many can be identified with our commonest species; others are more difficult of recognition under their somewhat lengthy but vague pre-Linnean designations.

In 1809, Professor Jameson read to the Wernerian Society of Edinburgh a paper, entitled, "A Catalogue of Animals of the Class Vermes, found in the Firth of Forth, and other parts of Scotland." Under the title Vermes he includes, as was usual at this date, representatives of the various invertebrate sub-kingdoms. Jameson enumerates seventy-seven species from the Firth, among which the only Molluscs are two species of *Tritonia*, one of *Doris* and one of *Chiton*. He divides the Vermes into—(1.) Mollusca, including Nudibranchs, Ascidians, Annelids, Holothurids, and Coelenterates; (2.) Testacea, of which *Chiton* is the only example given; (3.) Crustacea, including Echinids, Asterids, and Ophiurids; (4.) Coralla, being the Alcyonaria, Sponges, Polyzoa, and Sertularids; and (5.) Zoophyta, including *Pennatula*, *Hydra*, and *Coryne*.

In the same volume of the "Memoirs of the Wernerian Society," p. 370, is a paper by Captain Laskey, "An Account of North British Testacea," in which he enumerates between sixty and seventy species of Gastropods, Lamellibranchs, and Cirripedes, obtained in the Firth of Forth. Much doubt, however, has been expressed by distinguished conchologists as to the genuineness of many of the localities given in this paper.

Among those who have largely contributed to our knowledge of the fauna of the Forth, Mr H. D. S. Goodsir, who promised to attain to an equal scientific eminence with his brother, the illustrious anatomist, must be mentioned with special praise and regret. He discovered many new genera and species in the Forth, to which reference will be made in the text of this paper. He joined the Franklin Expedition in the capacity of naturalist, and shared in its disaster.

Many professors of the University of Edinburgh have from time to time investigated the fauna of the Forth, and of these Professor Jameson has already been mentioned; in more recent years Professors Goodsir, Edward Forbes, and Allman, held the most distinguished places.

Professor Goodsir was an anatomist in the widest and most philosophical signification of the word, and on this his great reputation is mainly based; but it may sometimes be forgotten that he was also an accomplished marine zoologist. Many of the species in our lists are given on his authority. In 1838, Professor Goodsir communicated to the Cupar Literary and Antiquarian Society a list of the marine animals collected at Anstruther by his brother, Harry Goodsir. We have not been able to ascertain whether this was ever published.

Professor Edward Forbes, although a pioneer in the investigation of marine faunas, does not seem to have personally done much dredging in the Forth. The period during which he occupied and adorned the Chair of Natural History in the University was too brief and too fully occupied with other work to admit of this.

The work of Professor Allman on the Hydroids and Polyzoa is well known, and his writings are indispensable to the student of marine zoology. They will be very frequently quoted in these lists.

The researches of Dr Thomas Strethill Wright are among the most valuable of those which we have to notice. Dr Wright gave much attention to the Protozoa and Coelenterata, and contributed to the Proceedings of the Royal Physical Society a series of papers, entitled, "Observations on British Zoophytes," in which the histology, physiology, and development of many of the Hydroids are treated in the most admirable manner.

About twenty years ago a committee of marine zoology was formed in the Royal Physical Society, for the purpose of dredging the Forth and neighbouring waters. The preparation of a list does not seem to have been an object with this committee, but some of their rarer finds are recorded in the Proceedings of the Society.

The most complete list hitherto published is that by Dr M'Bain, R.N., in the Rev. Walter Wood's book, "The East Neuk of Fife." The specimens mentioned in that work were mostly collected by Dr M'Bain himself and by Dr Howden, but some attention was also paid to the literature of the subject. The authority of this list will frequently be used in our paper.

We are especially indebted, and would now tender our best thanks, to Mr F. M. Balfour, F.R.S., Fellow and Lecturer of Trinity College, Cambridge, who has very liberally given us lists of the rarer forms which he has dredged in the Forth. Mr Balfour's assistance is the more valuable, as his work has principally been done in a part of the estuary which we have had comparatively little opportunity of investigating.

Dr M'Intosh's excellent "Marine Invertebrates and Fishes of St Andrews" should be consulted by students of the local fauna. From the proximity of the locality investigated, this work is very interesting in connection with the fauna of the Forth.

In addition to a study of the somewhat voluminous but scattered literature of the subject, our opportunities of acquiring a knowledge of the fauna of the Forth have consisted in—(1.) shore-collecting, extending over several years, at many points on both sides of the Firth; (2.) collecting from the refuse of the fishing boats at Newhaven and other piers; (3.) occasional night excursions with the oyster dredgers and long-line fishermen from Newhaven; and (4.) dredging excursions, in both sailing boats and steamers, organised in connection with the University Class of Practical Zoology.

ABBREVIATIONS USED IN THE LISTS.

G. J. A.,	Professor Allman, F.R.S.
F. M. B.,	F. M. Balfour, M.A., F.R.S.
Colds.,	Dr Coldstream.
Com. Mar. Zool.,	Committee of Marine Zoology of the Royal Physical Society.
J. G. D.,	Sir J. G. Dalyell.
Flem.,	Dr John Fleming.
E. F.,	Professor Edward Forbes.
H. D. S. G.,	Harry D. S. Goodsir.
J. G.,	Professor Goodsir.
Howd.,	Dr Howden.
R. J.,	Professor Jameson.
G. J.,	Dr George Johnston.

M'B.,	Dr M'Bain.
M. & B.,	Möbius and Butschli.
Br. Mus.,	Specimen from the Forth in the British Museum.
Ed. Mus.,	Specimen from the Forth in the Edinburgh Museum of Science and Art.
C. W. P.,	Charles W. Peach.
F. E. S.,	Professor Franz Eilhart Schulze.
Sim.,	Mr Simmons.
Th.,	Lieutenant Thomas, R.N.
T. S. W.,	Dr Thomas Strethill Wright.

COELENTERATA.

In the present part we can only give lists of the *Hydroida* and *Alcyonaria*, as the other sections of the Coelenterata, viz., the *Zoantharia*, the *Acalepha*, and the *Ctenophora* have not yet been sufficiently worked, and the material at our disposal is not extensive enough to enable us to compile anything like complete lists of these groups.

So far as the Hydroid Zoophytes are concerned, however, we have had plenty of material. They have always been a favourite group with marine zoologists, and have been studied in the Firth of Forth by successive generations of naturalists, who have frequently been rewarded by the discovery of species new to science.

Professor Allman and Dr Strethill Wright, whose researches we have already referred to in the Introduction, have contributed largely to our knowledge of the Hydroids. The section of Dr Wright's work most interesting to us, in connection with these lists, is his careful investigation of the more minute Zoophytes, among which he discovered so many new species—especially in the Athecata. Further work of the same description among the smaller Campanularians would almost certainly yield interesting and valuable results.

The nomenclature and arrangement in the following list are those given in Hinck's "History of the British Hydroid Zoophytes," from which we have in various ways derived the greatest assistance. For many of our names and localities we are indebted, in addition to the last-mentioned work and Dr Wright's papers, to Allman's "Gymnoblastic Hydroids," and various lists and notices by Professors Allman, Forbes, Jameson, Drs Fleming, M'Bain, George Johnston, F. E. Schulze, and others.

In our own investigations, comprising shore-collecting and dredging from fishing boats and steamers, extending over several years, we have taken the great majority of the species of Hydroids recorded as having been found in the Firth of Forth, and have been fortunate enough to discover several which have not hitherto been met with in this area. Our most notable deficiencies are among the minute Athecata discovered by Wright.

HYDROIDA.

I. ATHECATA—

CLAVIDÆ.

Clava multicornis (Försk.).
Firth of Forth (*T. S. W.*); Kincardine to Fifeness (*M'B.*); Kincardine, 2 fathoms (*Th.*); Firth of Forth (*G. J. A.*).
This species is common between tide marks on various parts of the shore. We have taken it at Joppa, at Wardie, and at South Queensferry.
It is the *Clava repens* of Wright (*Proc. Roy. Phys. Soc., Edin.*, 1857; *Ed. New Phil. Jour.*, July 1857).

C. squamata (Müll.).
Queensferry on *Fucus vesiculosus* (*T. S. W.*); Craigflower (*G. J. A.*).
In Dr Wright's paper on *Clava* this species figures as *C. membranacea* (*Ed. New Phil. Jour.*, July 1857).

C. nodosa (T. S. Wright).
Queensferry and Largo, on *Delesseria sanguinea* (*T. S. W.*).
This species was first described in *Proc. Roy. Phys. Soc., Edin.*, for 1862.

Turris neglecta (Lesson).
Queensferry (*T. S. W.*).

We found a few specimens some years ago near Joppa.

The gonozooid of this species was known for about twenty years before Wright traced the development of the fixed polypites, which he described under the name of *Clavula gossii* (*Ed. New Phil. Jour.*, July 1859).

HYDRACTINIIDÆ.

Hydractinia echinata (Fleming).
Firth of Forth (*T. S. W.*); common on univalves (*M'B.*); off the Bass Rock (*F. E. S.*).

This is a common species, and we have often dredged shells encrusted with it off Inchkeith, on the oyster bank, and in other parts of the Firth. Specimens may frequently be found on the shore after storms. F. E. Schulze* mentions having found this species on the beach between Portobello and Fisherrow.

PODOCORYNIDÆ.

Podocoryne carnea (Sars).
Inch Garvie (*G. J. A.*).
Clionistes reticulata (T. S. Wright).
Granton (*T. S. W.*).

CORYNIDÆ.

Coryne pusilla (Gaertn.).
Firth of Forth (*M'B.*).
Syncoryne eximia (Allman).
Firth of Forth (*G. J. A.*).
S. sarsii (Lovén.).
Firth of Forth (*T. S. W.*).

---

* II. Jahresh. d. Komm. z. Untersuch. d. deutsch. Meere in Kiel, p. 123. Berlin, 1875.

We took this species once in a pool near Wardie.

Mr F. M. Balfour informs us that he has dredged it in shallow water near the Bass Rock.

Syncoryne gravata (T. S. Wright).
North Berwick (*T. S. W.*).

S. decipiens (Dujard.).
Firth of Forth (*T. S. W.*).

S. ferox (T. S. Wright).
Firth of Forth (*T. S. W.*).

Gemmaria implexa (Alder).
Zanclea implexa, in "British Hydroid Zoophytes."
Inch Garvie (*T. S. W.*); Firth of Forth (*G. J. A.*).

Stauridium productum (T. S. Wright).
Caroline Park, on *Hydrallmania falcata* (*T. S. W.*).

EUDENDRIIDÆ.

Eudendrium rameum (Pall.).
Leith shore (*Colds.*); Firth of Forth (*Th.*); Firth of Forth (*J. G. D.*); Firth of Forth (*G. J. A.*).
We have obtained this species from the fishing boats at Newhaven.

E. ramosum (Linn.).
Firth of Forth (*M'B.*); Firth of Forth (*Th.*); Leith shore (*R. J.*); near the Bass Rock (*F. E. S.*); Firth of Forth (*Ed. Mus.*).
We have dredged this species frequently in from 3 to 8 fathoms in the Firth.

E. arbuscula (T. S. Wright).
Queensferry (*T. S. W.*).

E. capillare (Alder).
Firth of Forth, on *Delesseria sanguinea*, from 5 fathoms, and at low water on Ascidian's tests (*G. J. A.*).

ATRACTYLIDÆ.

Wrightia arenosa (Alder).
: *Atractylis arenosa,* in " British Hydroid Zoophytes."
: Largo (*T. S. W.*).

Perigonimus repens (T. S. Wright).
: Firth of Forth, on Sertularians, etc. (*T. S. W.*); Firth of Forth (*F. E. S.*).
: We dredged this species last summer near Inchkeith.

P. sessilis (T. S. Wright).
: On rocks at Granton, and on shells from deep water (*T. S. W.*).

P. palliatus (T. S. Wright).
: On a shell, Granton (*T. S. W.*).

P. vestitus (Allman).
: On a *Buccinum*, Firth of Forth (*G. J. A.*).

P. linearis (Alder).
: We dredged this species near Inchkeith last summer.

P. miniatus (T. S. Wright).
: On stones at Largo and at Granton (*T. S. W.*).

P. coccineus (T. S. Wright).
: Inch Garvie (*T. S. W.*).

P. bitentaculatus (T. S. Wright).
: Off Inchkeith (*T. S. W.*).

P. quadritentaculatus (T. S. Wright).
: Firth of Forth (*T. S. W.*).
: Mr Hincks and Professor Allman suspect that this and the preceding species may both prove immature forms.

Garveia nutans (T. S. Wright).
: Inch Garvie (*T. S. W.*); Firth of Forth on Algæ, etc. (*G. J. A.*).
: Described by Professor Allman as *Eudendrium bacciferum* (*Ann. N. H.*, July 1859).

Bimeria vestita (T. S. Wright).
>Bimer Rock, North Queensferry, and Inch Garvie (*T. S. W.*); Firth of Forth on Algæ, etc. (*G. J. A.*).

Dicoryne conferta (Alder).
>Firth of Forth (*G. J. A.*).
>We dredged this species on *Apporhais pes-pelicani* off Kirkcaldy last summer.

Bougainvillea ramosa (v. Ben.).
>Queensferry (*T. S. W.*).
>We have taken it in 7 fathoms.

B. fruticosa (Allman).
>Firth of Forth (*G. J. A.*).

Bougainvillea sp.
>Some specimens of *Bougainvillea* dredged last summer off Inchkeith we are unable to identify with any of the known species. In some respects they resemble the variety of *B. muscus* mentioned by Hincks on page 112 of the "British Hydroid Zoophytes."

TUBULARIIDÆ.

Tubularia indivisa (Linn.).
>Firth of Forth (*M'B.*); Firth of Forth (*Ed. Mus.*).
>We have dredged this common species from 7 fathoms in the Firth.

T. larynx (Ell. and Sol.).
>Firth of Forth (*M'B.*); Kincardine (*Th.*); Firth of Forth (*Ed. Mus.*); Firth of Forth (*F. E. S.*).

T. coronata (Abildg.).
>Firth of Forth (*M'B.*).
>We have dredged this species from 10 fathoms.

T. attenuata (Allman).
>Firth of Forth, 15 fathoms (*G. J. A.*).

Corymorpha nutans (Sars).
>Firth of Forth, 14 fathoms (*G. J. A.*).

PENNARIIDÆ.

Vorticlava proteus (T. S. Wright).
Fluke Hole (*T. S. W.*).

II. THECAPHORA—

CAMPANULARIIDÆ.

Clytia johnstoni (Alder).
Firth of Forth (*F. E. S.*).
This species, the *Campanularia volubilis* of Johnston, is common in the Firth of Forth, chiefly on other zoophytes; we have taken it frequently.

Obelia geniculata (Linn.).
Firth of Forth (*M'B.*); Newhaven (*Ed. Mus.*).
In profusion on *Laminaria* all along the coast; we have taken it at Wardie, Aberdour, Elie, etc.

O. longissima (Pall.).
We found this species on the shore at Longniddry, cast up after a storm, some years ago. We also dredged it in the Firth of Forth last summer.

O. dichotoma (Linn.).
On the beach between Portobello and Fisherrow (*F. E. S.*); Firth of Forth (*Ed. Mus.*); Firth of Forth (*G. J.*); Firth of Forth (*Th.*); off the Bass Rock (*F. E. S.*).
We dredged this species last summer between Inchkeith and Kirkcaldy.

Campanularia volubilis (Linn.).
Firth of Forth (*Th.*); Firth of Forth (*M'B.*); Firth of Forth (*Ed. Mus.*).
We have obtained this species from 5 fathoms.
It is not improbable that the species referred to under this name by M'Bain was *Clytia johnstoni* (Alder).

Campanularia verticillata (Linn.).
　　　　　　Firth of Forth (*Th.*); Firth of Forth (*M'B.*).
　　　　　　Not common. We have dredged it from about 10 fathoms.

C. flexuosa (Hincks).
　　　　　　Firth of Forth (*M'B.*).
　　　　　　Dredged last summer near Kirkcaldy.

C. decipiens (T. S. Wright).
　　　　　　Firth of Forth (*T. S. W.*).

C. integra (Macgill.) ?
　　　　　　Dredged last summer near Inchkeith.

Thaumantias inconspicua (Forb.).
　　　　　　Firth of Forth (*T. S. W.*).
　　　　　　This species was only known by the gonozooid until Dr Strethill Wright reared the polypites from the ova.

Gonothyræa lovéni (Allman).
　　　　　　Cramond Island (*G. J. A.*).
　　　　　　We have taken this species in considerable quantity about low water mark at Wardie.

CAMPANULINIDÆ.

Campanulina acuminata (Alder).
　　　　　　Firth of Forth (*T. S. W.*).

C. repens (Allman).
　　　　　　On Sertularians from 5 fathoms (*G. J. A.*).

Opercularella lacerata (Johnst.).
　　　　　　Morrison's Haven (*T. S. W.*).
　　　　　　We found this species a few years ago at Newhaven.

LAFOEIDÆ.

Lafoëa dumosa (Fleming).
　　　　　　Firth of Forth (*M'B.*); Firth of Forth (*Ed. Mus.*); Firth of Forth (*F. E. S.*).
　　　　　　Common in the Firth, on the oyster bank, in 3 to 10 fathoms; usually parasitic on

other zoophytes, such as *Hydrallmania falcata.*

Lafoëa fruticosa (Sars).
Firth of Forth (*F. E. S.*).
We have dredged this species once or twice along with the preceding.

Calycella syringa (Linn.).
Off the Bass, Firth of Forth (*F. E. S.*).
Not uncommon, on other zoophytes.

Filellum serpens (Hassall).
The *Reticularia immersa* of Sir Wyville Thomson (*Ann. Mag. N. H.*, 2d ser., vol. xi., p. 443).
We have obtained this species from the refuse in the Newhaven fishing boats; it was also dredged last summer near Inchkeith.

TRICHYDRIDÆ.

Trichydra pudica (T. S. Wright).
Fluke Hole (*T. S. W.*).

Coppinia arcta (Dalyell).
We have found this species frequently in the fishing boats at Newhaven.

HALECIIDÆ.

Halecium halecinum (Linn.).
Firth of Forth (*M‘B.*); Firth of Forth (*Th.*); Firth of Forth (*F. E. S.*); Firth of Forth (*Ed. Mus.*).
One of the commonest species in the Firth; very common on the oyster bank; dredged frequently from 5 fathoms.

H. muricatum (Ell. and Sol.).
Firth of Forth (*C. W. P.*); Firth of Forth (*R. J.*); Firth of Forth (*M‘B.*); Firth of Forth (*Ed. Mus.*).
We obtained a fine specimen a few years

ago from the Newhaven fishing boats. We have also collected it at North Berwick, cast up after a storm.

Halecium plumosum (Hincks)?
Dredged last summer.

H. beanii (Johnst.).
Queensferry (*T. S. W.*); Firth of Forth (*Ed. Mus.*).
We have dredged this species in 8 fathoms between Inchkeith and Kirkcaldy.

SERTULARIIDÆ.

Sertularella polyzonias (Linn.).
Firth of Forth (*M'B.*).
A very common species in about 5 fathoms on the oyster bank westwards from Inchkeith, where it attains a large size.

S. rugosa (Linn.).
Firth of Forth (*M'B.*).
We obtained this species frequently at low water mark at Elie some years ago.

Diphasia rosacea (Linn.).
Firth of Forth (*M'B.*); Firth of Forth (*F. E. S.*).
Not rare on shells and stones from a few fathoms in depth. We dredged it frequently last summer near Inchkeith.

D. fallax (Johnst.).
Firth of Forth, plentiful (*Colds.*); Firth of Forth (*M'B.*).

D. attenuata (Hincks).
Firth of Forth, in 22 fathoms on a sandy and shelly bottom (*F. E. S.*).

D. tamarisca (Linn.).
Firth of Forth (*Colds.*); Firth of Forth (*M'B.*); Firth of Forth (*Ed. Mus.*).
We dredged this species last summer near the Isle of May.

Sertularia pumila (Linn.).
> Firth of Forth (*M'B.*); Firth of Forth (*Ed. Mus.*).
> Common between tide marks on stones, *Fucus*, etc.; in profusion at Wardie.

S. operculata (Linn.).
> Firth of Forth (*M'B.*); Firth of Forth (*Colds.*).
> We have found large masses of this species cast up on the shore between Longniddry and North Berwick.

S. filicula (Ell. and Sol.).
> Firth of Forth (*M'B.*).
> We have dredged this species frequently. We got it last summer in 7 fathoms near Inchkeith.

S. abietina (Linn.).
> Firth of Forth (*M'B.*); Firth of Forth (*Ed. Mus.*).
> Not uncommon on stones and shells from the oyster bank.

S. argentea (Ell. and Sol.).
> Firth of Forth (*M'B.*); Firth of Forth, 8 fathoms (*F. E. S.*).
> This species is not uncommon. We took it frequently last summer on old shells, etc., from between Inchkeith and Kirkcaldy.

S. cupressina (Linn.).
> Firth of Forth (*M'B.*); Firth of Forth (*R. J.*); Firth of Forth (*Th.*); Firth of Forth (*Ed. Mus.*).
> We have dredged this species several times in the Firth.

Hydrallmania falcata (Linn.).
> Firth of Forth (*M'B.*); off the Bass Rock (*F. E. S.*).
> Probably the commonest Hydroid in a few fathoms of water off Newhaven. It occurs

in profusion in some places, and attains a
large size.

Thuiaria thuja (Linn.).
Firth of Forth (*M'B.*); near Inchkeith
(*Th.*); Leith shore (*R. J.*); Firth of Forth
(*Ed. Mus.*); Firth of Forth (*F. E. S.*).
Common in 5 fathoms and upwards.

T. articulata (Pallas).
We dredged this species near Inchkeith
last summer.

PLUMULARIIDÆ.

Antennularia antennina (Linn.).
Firth of Forth (*M'B.*).
We have dredged this species several
times. It is sometimes found cast up on
the beach.

A. ramosa (Lamk.).
Firth of Forth (*M'B.*); Firth of Forth,
in 25 fathoms (*F. E. S.*).
We dredged this species last summer in
10 fathoms. It is not nearly so common
in the Firth of Forth as *A. antennina*, but
at Lamlash, on the west coast, it occurs in
profusion, and is much the commoner of
the two species.

Plumularia pinnata (Linn.).
Firth of Forth (*M'B.*); Firth of Forth
(*Ed. Mus.*); Firth of Forth, 25 fathoms
(*F. E. S.*).
Not uncommon; dredged last summer
between Inchkeith and Kirkcaldy.

P. setacea (Ellis).
Firth of Forth (*M'B.*).
We obtained this species plentifully at
low water mark near Elie some years ago.

P. catharina (Johnst.).
Firth of Forth (*M'B.*); Firth of Forth
(*Colds.*); Firth of Forth, on a sandy and

B

shelly bottom in 30 and in 22 fathoms (*F. E. S.*).

We dredged this species on several occasions last summer from about 10 fathoms.

ALCYONARIA.

ALCYONIDÆ.

Alcyonium digitatum (Linn.).

Firth of Forth (*F. E. S.*).

Very common on the oyster bank. We have obtained it between tide marks attached to the rocks, near Elie.

GORGONIADÆ.

Gorgonia flabellum (Linn.).

Professor Jameson ('Wernerian Memoirs,' vol. i., p. 561) recorded this species as having been found by Mr Mackay on Leith shore.

Professor Goodsir dredged a large specimen in the Forth.

A young friend, Mr Malcolm Laurie, lately picked up a worn specimen on Portobello beach.

All the specimens seem to have been dead; and there can be little doubt that the species never occurs in a fresh condition in these seas.

PENNATULIDÆ.

Pennatula phosphorea (Linn.).

Firth of Forth (*F. E. S.*); Firth of Forth (*Ed. Mus.*).

We have obtained this species cast ashore, and in the fishing boats at Newhaven.

Virgularia mirabilis (Linn.).

Dredged near Inchkeith (*Sim.*); Prestonpans Bay (*R. J.*); Firth of Forth (*Ed. Mus.*).

ECHINODERMATA.

In this list the Crinoidea are omitted, as no specimen of this group has hitherto been found in the Firth. *Antedon rosaceus*, which is extremely common on the west coast of Scotland, may yet be found within our limits, as its area of distribution seems to extend round the northern coast; and Mr C. W. Peach has recorded it off Peterhead. *Antedon sarsii*, a northern species, may be looked for in the seaward limit of the Forth. In the list of Ophiuroidea the arrangement and nomenclature of Professor Lyman, as given in his "Ophiuridæ and Astrophytidæ,"* has been followed; and in the Asteroidea, Professor Ed. Perrier's arrangement, used in his "Révision de la Collection de Stellérides du Muséum d'Histoire Naturelle de Paris."† For the Echinoidea, we have followed Professor Alexander Agassiz' nomenclature, as given in his "Revision of the Echini."‡

Much work must still be done in the investigation of the Holothuroidea of the British Seas before they are properly understood, and the list of the Forth species of this group, which we give, is that which we now regard with least satisfaction. Although from time to time a considerable number of species of Holothurids have been obtained in the Forth, the group is very poor in individuals, so that specimens for comparison are rarely obtained.

We have obtained the greatest assistance in the preparation of this list of Echinodermata from the works above mentioned. Also from Professor Edward Forbes' "History of British Starfishes," and from the Rev. A. M. Norman's valuable paper on the "Crinoidea, Ophiuroidea, and Asteroidea of the British Seas."§ Although not always adopting the nomenclature of Professor Forbes, we have in every case given the names used by him in his description of Echinoderms, as his work is in the hands of every student of the group.

---

* Mem. Mus. Comp. Zool., Cambridge, Mass., 1864.
† Archives de Zoologie Expérimentale, 1875.
‡ Mem. Mus. Comp. Zool., Cambridge, Mass., 1874.
§ Ann. and Mag. Nat. Hist., 1865, p. 98.

OPHIUROIDEA.

OPHIURIDÆ.

Ophioglypha lacertosa (Linck).
Firth of Forth (*E. F.*); Firth of Forth (*M'B.*).
This species is the *Ophiura texturata* of Forbes. It is common near the mouth of the Firth, and becomes scarcer higher up. We have dredged it on several occasions at depths of 4 to 7 fathoms.

O. albida (Forbes).
Firth of Forth (*E. F.*); Firth of Forth (*M'B.*).
The *Ophiura albida* of Forbes. It is a very abundant species in the Firth, and we always obtained it in great numbers on the oyster banks near Inchkeith, and further down the estuary. It prefers sandy bottoms.

O. affinis (Lütken).
Bass Rock, 24 fathoms (*M. & B.*).*

Ophiocoma nigra (O. F. Müll.).
Firth of Forth (*M'B.*); Firth of Forth (*F. M. B.*). On the deep-sea lines.
Ophiocoma granulata of Forbes. This species is not common in the Firth. We have obtained it at Newhaven, the specimens having probably been brought in on the fishermen's lines from the vicinity of the May Island. On some parts of the west of Scotland this is the most abundant species in a few fathoms of water.

Ophiopholis bellis (Johns.).
Firth of Forth (*Howd.*); Prestonpans (*Ed. Mus.*); Firth of Forth (*F. M. B.*).

---

* II. Jahresb. d. Komm. z. Untersuch. d. deutsch. Meere in Kiel, IV., Echinodermata.

The *Ophiocoma bellis* of Forbes. This beautiful species is somewhat sparsely distributed in the Forth. We have taken it on the oyster banks at 5 fathoms, and in deeper water in Aberlady Bay, and have obtained it while shore-collecting near the piers.

Amphiura filiformis (Müll. sp.).

Firth of Forth (*Howd.*); Bass Rock, 24 fathoms (*M. & B.*).

The *Ophiocoma filiformis* of Forbes. This species is found on the west coast of Scotland and in the Irish seas. Mr F. M. Balfour found it plentiful on mud and gravel near Dunbar.

A. chiagii (Forbes).

Off Anstruther (*H. D. S. G.*); Firth of Forth (*F. M. B.*); Bass Rock, 24 fathoms (*M. & B.*).

The first record of the occurrence of this species in the Forth is that of a specimen obtained from the stomach of a cod off Anstruther, by Mr H. Goodsir, and which was described as *Ophiocoma punctata* by Forbes in his "History of British Starfishes." In 1845, Forbes described an ophiurid [*] (*Amphiura chiagii*), which is supposed by the Rev. A. M. Norman to be the adult form of Forbes' earlier *O. punctata*. *Amphiura chiagii* is not uncommon near the mouth of the Forth, and is found on both the western and eastern coasts.

A. squamata (Dell. Ch. sp.).

Newhaven (*J. G.*); Firth of Forth (*Howd.*); Dunbar (*Ed. Mus.*).

The *Ophiocoma neglecta* of Forbes. We

---

[*] Trans. Linn. Soc., vol. xix., p. 151.

have often found it under stones and in rock pools while shore-collecting. It is with us nearly confined to the littoral and upper laminarian zones.

Ophiocnida ballii (Thompson).

Off Anstruther (*J. G.*).

In his "History of British Starfishes," Forbes mentions two species of Ophiocoma, viz., *Ophiocoma Ballii* of Thompson, from the coast of Dublin, and a new species, *O. Goodsiri*, from the Forth. From a comparison of type specimens of these, Mr Norman has determined that *O. Ballii* and *O. Goodsiri* belong to the same species, and of the two specific names, that of *Ballii* has the priority.

Amphiura ballii is found on the northern and eastern coasts, and in the Irish Seas.

O. brachiata (Mont.).

Firth of Forth (*Howd.*).

The *Ophiocoma brachiata* of Forbes. We have never found it in the Forth, and include it in our list on the authority of Dr Howden. *O. brachiata* has elsewhere been recorded from the western and Irish coasts.

Ophiothrix rosula (Linck).

Leith (*Ed. Mus.*); Firth of Forth (*M'B.*); Firth of Forth, 22 fathoms (*M. & B.*).

The *Ophiocoma rosula* of Forbes. This is the most abundant of the Forth species of Ophiuridæ, occurring in great numbers on the oyster and shell banks, and extending into deep water. The pigmentation of different individuals is very varied, but it is always easily recognised by the generic character of the two large triangular plates on the dorsal surface opposite the origin of each ray.

ASTROPHYTIDÆ.

Astrophyton linckii (M. and T.).

We include this species in our list with the most complete reservation, our only authority being a manuscript of Captain Laskey's quoted by Forbes, in which he mentions having obtained "a great Medusa's-head Starfish in a herring net at Dunbar." *A. linckii* is a northern species, and was frequently dredged during the "Porcupine" Exploring Expedition. The probability of its occurrence in the Forth is very slight. Mr Norman, for whose opinion we have the highest respect, considers that its area of distribution is confined to the seas of Scandinavia and Shetland.

ASTEROIDEA.

ASTERIADÆ.

Asterias rubens (Linn.).

Firth of Forth (*Ed. Mus.*).

The *Uraster rubens* of Forbes. This is the most abundant species of Asterid in the Forth, being brought up in numbers at almost every dredging. It is especially common on the shell banks at about 5 fathoms, but we also found it at all the localities which we have dredged. Forbes describes a variety *A. coriacea* from the Forth, which is characterised by the prominent spines of the dorsal ridge. We have frequently obtained this variety.

A. violacea (O. F. Müll.).

Firth of Forth (*E. F.*); Firth of Forth (*Howd.*).

The *Uraster violacea* of Forbes. We have often obtained this species or variety in the Firth, and have always been inclined to regard it as a mere variety of the widely-distributed and polymorphic *Asterias rubens*. At present, however, we have followed Mr Norman and Professor Perrier in giving it the rank of a separate species. Forbes states that it is by far commoner than *A. rubens* at the mouth of the Firth.

ECHINASTERIDÆ.

Solaster papposus (Linn.).

Firth of Forth (*Ed. Mus.*).

Solaster papposa of Forbes. It is a common species in the Firth. We often got it at low tide, and have dredged it from shallow water to 14 fathoms. It attains a very large size, and in this respect offers a marked contrast to its dwarfed northern congeners, *S. borealis* and *S. furcifer*.

S. endeca (Linn.).

Anstruther (*J. G.*); Firth of Forth (*M'B.*). This species is less common than *S. papposus*. We never obtained it at low water on the rocks, but dredged it frequently at depths of 5 to 16 fathoms.

Cribrella oculata (Linck).

Firth of Forth (*M'B.*); Firth of Forth (*F. M. B.*).

The *Cribella oculata* of Forbes. It is a not uncommon species. We have found it among rocks at low water at Aberdour and near Newhaven, and have dredged it on the oyster banks and elsewhere at greater depths.

GONIASTERIDÆ.

Hippasteria plana (Linck sp.).

Firth of Forth (*Dr Neill*);[*] Firth of Forth (*F. M. B.*).

[*] Fleming, "Hist. Brit. Animals," p. 486. Edin., 1828.

This species is the *Goniaster equestris* and *G. abbensis* of Forbes. It must be regarded as an inhabitant of only the seaward limits of the Firth. Mr F. M. Balfour has obtained one specimen from the Forth, which he thinks was from the deep-sea lines, or possibly from the creels. In Mr Norman's account of its general distribution, it is said to occur on the western, evidently a misprint for the eastern, coast of North America. It is abundant on the cod banks off Halifax.

ASTROPECTINIDÆ.

Luidia savignyi (Aud.).

Firth of Forth (*Howd.*).
Professor Perrier includes under this specific name the *Luidia fragilissima* of Forbes = *Asterias savignyi* of Audouin and *Luidia sarsii* of Düben and Koren; and in this we follow him. Although it has been obtained on several occasions near the mouth of the estuary, we have not been so fortunate as to find it. Our friend Dr Traquair informs us that he has dredged it off North Berwick.

Astropecten irregularis (Linck).

Firth of Forth (*E. F.*); Firth of Forth (*M'B.*). The *Asterias aurantiaca* of Forbes, according to whom it is common in the Forth. We have found it on several occasions among the refuse of the fishing lines, and while dredging in rather deep water; but it can scarcely be reckoned one of our commoner species, at least in the upper reaches of the estuary. Mr F. M. Balfour has found it rather commonly in the creels, and while dredging near the mouth of the Firth.

ECHINOIDEA.

Echinidæ.

Echinus esculentus (Penn. sp.).
: Firth of Forth (*E. F.*); Firth of Forth (*M'B.*).
We have followed Professor Alex. Agassiz' nomenclature of this species, which is the *E. sphaera* of Forbes. It is our most common sea-urchin, and occurs from between tide marks to the greatest depths in the Firth. We often dredged it on the banks around Inchkeith, and at various other places.

E. miliaris (Linn.).
: Firth of Forth (*Ed. Mus.*).
We have never found this species in the littoral zone, but have obtained it at 5 fathoms off Inchkeith, and in the north channel in 18 fathoms.

Strongylocentrotus dröbachiensis (O. F. Müll.).
: The *Echinus neglecta* of Forbes. Mr F. M. Balfour has found this species on a muddy bottom, at 30 fathoms, near the mouth of the estuary.

Clypeastridæ.

Echinocyamus pusillus (Müll.).
: We found some dead specimens on the shore at Largo, and have dredged it. Mr F. M. Balfour found it common near Dunbar.

Spatangidæ.

Spatangus purpureus (Leske).
: Leith shore (*R. J.*); Firth of Forth (*Flem.*); Firth of Forth (*M'B.*).
Not uncommon. We have frequently dredged it in 5 to 12 fathoms.

Echinocardium cordatum (Penn.).
: Firth of Forth (*M'B.*).
: The *Amphidotus cordatus* of Forbes. We have obtained it at Elie, and in other sandy bays on the shores of the Firth. It is often cast on shore in great numbers after storms.

E. flavescens (Müll.).
: Leith sands (*Colds.*); Bass Rock, 24 fathoms (*M. & B.*).
: The *Amphidotus roseus* of Forbes. The records of its occurrence in the Forth, in the "History of British Starfishes," and that by Möbius and Butschli, are the only ones we possess.

HOLOTHUROIDEA.

PSOLIDÆ.

Psolus phantapus (Linn.).
: Coast of Fife (*J. G.*); on a baited line near Inchkeith (W. S. Young, *Proc. Roy. Phys. Soc.*, vol. ii.).

PENTACTÆ.

Cucumaria frondosa (Gunner).
: Off coast of Fife (*J. G.*).
: This is a northern species, and was obtained in great numbers in Lerwick Bay during the "Porcupine" Expedition. We have not obtained it in the Forth, but it occurs as an inhabitant of the seaward limits. Professor R. O. Cunningham informs us that he has got it on two occasions from the Prestonpans fishermen, who obtained it near the mouth of the Firth.

C. elongata (Düben and Koren).
: Bass Rock, 24 fathoms (*M. & B.*).
: Occurs at the mouth of the Firth, and is occasionally obtained by the trawlers and

on the deep-sea lines. The *Holothuria montaguii*, founded by Dr Fleming on a specimen obtained in the Forth by Dr Neill, was probably this species.

Cucumaria lactea (Forb. and Goods.).

Off Fife coast (*J. G.*); Firth of Forth (*M'B.*); Firth of Forth, 22 fathoms (*M. & B.*).

This is one of the commoner species near the mouth of the Firth. Mr D'Arcy Thompson informs us that he has obtained several specimens.

Thyonidium commune (Forb. and Goods.).

East coast of Fife (*J. G.*).

The *Cucumaria communis* of Forbes. We obtained one specimen from the stomach of a cod.

Thyone fusus (Müll.).

Anstruther (*J. G.*); Firth of Forth (*Com. Mar. Zool.*).

The *Thyone papillosa* of Forbes. Mr F. M. Balfour informs us that he has taken this species at 12 to 16 fathoms on mud and shells off North Berwick.

SYNAPTIDÆ.

Synapta inhærens (Müll.).

Bass Rock, 24 fathoms (*M. & B.*).

Mr F. M. Balfour has found this species under stones, between tide marks, at the mouth of the Tyne, near Dunbar.

In addition to these, two species are described by Dr Fleming in his "History of British Animals," under the names of *Holothuria dissimilis* = *Cucumaria dissimilis*, Forbes (Leith shore, *Colds.*), and *Holothuria neillii* = *Cucumaria neillii*, Forbes (Firth of Forth, *Dr Neill*).

PROTOZOA.

In this list the Foraminifera only are catalogued, the other Protozoan groups being omitted for the present, as they require very special work. With the exception of a few species of Infusoria and Lobosa, noted by Dr Strethill Wright, they have been entirely neglected by workers in the Forth. The Sponges, now considered as a group of the Metazoa, will be treated separately.

In the preparation of the following list we were very largely indebted to a valuable paper by Mr H. B. Brady, F.R.S., on "Brackish-Water Foraminifera,"* one of the localities which he investigated being the upper basin of the Forth. We have also used the list published by Professor Franz Eilhart Schulze, in the report of the German Expedition of 1872.†

In the identification of the forms which we have ourselves obtained in the Forth, we have had the kind assistance of Mr Frederick Pearcey, to whom we would now express our thanks. Our material was dredged in 5 fathoms, on the oyster bank, near Inchkeith. We have added a number of forms to those previously recorded from the Forth, but do not doubt that further work will very largely increase the list. Mr Pearcey reports that the gathering in which the Foraminifera were found consisted of quartz grains, small and broken shells, Annelid tubes, and fragments of Polyzoa, Hydroids, and Echinoderms. Mr H. B. Brady's specimens were obtained about 20 yards from the pier at Bo'ness, in 3 to 4 fathoms, the bottom consisting of soft, slimy mud. Those in Franz Eilhart Schulze's list were got in the beach sand between Portobello and Fisherrow, and near the Bass Rock.

Mr Brady's specimens are marked (*H. B. B.*).

FORAMINIFERA.

I. IMPERFORATA—

MILIOLIDÆ.

Quinqueloculina seminulum (Linn.).

Off pier, Bo'ness (*H. B. B.*); beach sand be-

---

* Ann. and Mag. Nat. Hist., 1870, p. 273.

† II. Jahresb. d. Komm. z. Untersuch. d. deutsch. Meere in Kiel, I., Protozoa. Berlin, 1874.

tween Portobello and Fisherrow (*F. E. S.*); off Bass Rock, 24 fathoms (*F. E. S.*).

We found this species very abundant on the oyster bank off Inchkeith.

Quinqueloculina subrotunda (Mont.).

Off pier, Bo'ness (*H. B. B.*); beach sand between Portobello and Fisherrow (*F. E. S.*).

Spiroloculina limbata? (D'Orb.).

We obtained one dead specimen of *Spiroloculina* from 5 fathoms, off Inchkeith. It was somewhat damaged, so that the specific characters could not with certainty be made out.

LITUOLIDÆ.

Lituola scorpiurus (Montfort).

Mr H. B. Brady found a single specimen of the stunted form of this species off Bo'ness, and did not obtain it from any other of the localities which he investigated for brackish-water Foraminifera.

L. canariensis (D'Orb.).

Off pier, Bo'ness (*H. B. B.*).

We found many specimens off Inchkeith.

II. PERFORATA—

LAGENIDÆ.

Lagena sulcata (W. and J.).

Off pier, Bo'ness (*H. B. B.*).

L. lyellii (Sequenza).

Off pier, Bo'ness (*H. B. B.*).

L. lævis (Mont.).

Off pier, Bo'ness (*H. B. B.*).

L. gracillima (Sequenza).

Off pier, Bo'ness (*H. B. B.*).

L. globosa (Mont.).

Off pier, Bo'ness (*H. B. B.*).

L. striata (D'Orb.).

Off pier, Bo'ness (*H. B. B.*); beach sand between Portobello and Fisherrow (*F. E. S.*).

Lagena distoma (P. and J.).
　　Off pier, Bo'ness (*H. B. B.*).
L. marginata (W. and J.).
　　Off pier, Bo'ness (*H. B. B.*).
　　Mr Brady also found at Bo'ness the subvarietal form *Entosolenia marginata*, var. *lucida* (Williamson). We obtained the typical *L. marginata* off Inchkeith.
L. squamosa (Mont.).
　　A few specimens were found in the material from the oyster bank, 5 fathoms.
Nodosaria scalaris (Batsch).
　　Off pier, Bo'ness (*H. B. B.*).
Dentalina communis (D'Orb.).
　　Off pier, Bo'ness (*H. B. B.*).
D. guttifera (D'Orb.).
　　Off pier, Bo'ness (*H. B. B.*).
Polymorphina lactea (W. and J.).
　　We obtained this species off Inchkeith. Mr Brady found it in only two of his thirty-two brackish-water localities.

GLOBIGERINIDÆ.

Globigerina bulloides (D'Orb.).
　　We found many specimens of this widely distributed species off Inchkeith in 5 fathoms. It was not obtained by Mr Brady at Bo'ness, which is about fourteen miles further up the estuary than our locality.
Textularia sagittula (Defrance).
　　Plentiful off Inchkeith, 5 fathoms.
Verneuilina polystropha (Reuss).
　　Off Inchkeith, this species was fairly plentiful.
Bulimina marginata (D'Orb.).
　　Off pier, Bo'ness (*H. B. B.*).
B. ovata (D'Orb.).
　　Off pier, Bo'ness (*H. B. B.*); beach sand

between Portobello and Fisherrow (*F. E. S.*); Bass Rock, 24 fathoms (*F. E. S.*).

Bulimina elegantissima (D'Orb.).
Off pier, Bo'ness (*H. B. B.*).

Bolivina punctata (D'Orb.).
Off pier, Bo'ness (*H. B. B.*).

B. plicata (D'Orb.).
Off pier, Bo'ness (*H. B. B.*).

Discorbina globularis (D'Orb.).
Off pier, Bo'ness (*H. B. B.*); abundant, attached to Polyzoa, Hydroids, stones, etc., off Inchkeith.

D. rosacea (D'Orb.).
We found a few specimens on the oyster bank.

Planorbulina mediterranensis (D'Orb.).
Off pier, Bo'ness (*H. B. B.*).
We found a broken specimen of *Planorbulina*, which we have not been able to identify, on the oyster bank.

Truncatulina lobatula (Walker).
Off pier, Bo'ness (*H. B. B.*).

Rotalia beccarii (Linn.).
Off pier, Bo'ness (*H. B. B.*); near Bass Rock (*F. E. S.*).
This is a very abundant species in the Forth. We have obtained it repeatedly, and from many localities. In the material from the oyster bank it is very plentiful.

Tinoporus lævis (Parker and Jones, sp.).
Mr Brady obtained a single worn specimen in his material from Bo'ness. He thinks that it was probably a dead shell carried in by the tide from the deeper sea.

NUMMULINIDÆ.

Polystomella striato-punctata (F. and M.).
Very plentiful off pier, Bo'ness (*H. B. B.*);

beach sand between Portobello and Fisherrow (*F. E. S.*).

Nonionina depressula (W. and J.).

Off pier, Bo'ness (*H. B. B.*); beach sand
between Portobello and Fisherrow (*F.E.S.*).

N. umbilicatula (Mont.).

Beach sand between Portobello and Fisherrow (*F. E. S.*).

POLYZOA.

Polyzoa are abundant in the Firth of Forth, in the coralline zone, and especially on the oyster bank, where some species occur in profusion, and attain a large size. Notwithstanding this richness in specimens, the number of species hitherto recorded from the Firth is small, and the present list is probably far from complete. We believe that many of the smaller species will yet be found here, especially those formerly united under the generic title *Lepralia* (as used by Johnston in the "British Zoophytes").

We have followed the nomenclature and arrangement given in Hincks' recent work on the British Polyzoa.* From this exhaustive treatise we have derived much valuable assistance, especially in the identification of species and the determination of their synonymy.

We would tender our thanks to the veteran and well-known marine zoologist Mr C. W. Peach, to whom we are indebted for a list of the rarer species of Polyzoa which he has obtained from the Firth of Forth.

Several of our species are recorded on the authority of Dr Kirchenpauer's report on the Polyzoa collected by the German Expedition of 1872, which investigated the fauna of the North Sea.† These are marked (*K.*).

Dr M'Bain's lists in "The East Neuk of Fife" have again contributed largely to the number of the recorded species.

---

* "A History of the British Marine Polyzoa." By Thomas Hincks, B.A., F.R.S. London, Van Voorst, 1880.

† II. Jahresb. d. Komm. z. Untersuch. d. deutsch. Meere in Kiel, VI., Bryozoen. Berlin, 1874.

CHEILOSTOMATA.

EUCRATIIDÆ.

Eucratea chelata (Linn.).
: Among Zoophytes from the Firth of Forth.

Gemellaria loricata (Linn.).
: This is one of the commonest species of Polyzoa in the Firth of Forth. It is recorded by Kirchenpauer from 24 fathoms near the Bass Rock, and we have dredged it in 26 fathoms, off the Isle of May. It occurs over almost the entire area, and in profusion on the oyster bank. It is often found washed up on the shore.

CELLULARIIDÆ.

Cellularia peachi (Busk).
: On the long-lines, Newhaven, very rare (*C. W. P.*); Firth of Forth, under the name of *Cellularia neritina* (*M'B.*).

Menipea ternata (Ell. and Sol.).
: We dredged this species last summer in the Firth of Forth.

Scrupocellaria scruposa (Linn.).
: Firth of Forth (*M'B.*).
: Common in a few fathoms of water.

S. scabra (v. Ben.).
: Firth of Forth, one specimen (*C. W. P.*).

S. reptans (Linn.).
: Firth of Forth (*M'B.*).
: The *Canda reptans* of various authors.
: We dredged this species several times last summer, and have also found it among the refuse from the dredges on Newhaven pier.

BICELLARIIDÆ.

Bicellaria ciliata (Linn.).
: On roots of *Fuci* (*M'B.*); Firth of Forth (*C. W. P.*); off Aberdour, 5 fathoms (*nob.*).

We have dredged this species several times, last summer and previously.

Bugula avicularia (Linn.).
Firth of Forth (*M'B.*).
We dredged this species last summer to the west of Inchkeith in 5 fathoms.

B. plumosa (Pallas).
Dredged last summer.

B. murrayana (Johnst.).
Leith and Newhaven, on the fishermen's nets (*D. Landsborough, jun.*).

CELLARIIDÆ.

Cellaria fistulosa (Linn.).
This species is common in the Forth. We have dredged it frequently on the oyster bank, and took it last summer in 14 fathoms, off Longniddry.
The *Salicornaria farciminoides* of Johnston and others.

FLUSTRIDÆ.

Flustra foliacea (Linn.).
Common on the oyster bank. We have taken it lately from 12 fathoms, north-east of Inchkeith; 5 fathoms, west of Inchkeith; and 4 fathoms, off Aberdour.

F. securifrons (Pallas).
The *Flustra truncata* (Linn.) of Johnston, Busk, and others. Leith, Dirleton, and opposite the Bass Rock (*D. Landsborough, jun.*); Portobello and Firth of Forth, 30 fathoms (*K.*); Firth of Forth (*M'B.*).
This is a fairly common species in the Forth. We have dredged it west of Inchkeith in 5 fathoms, and north-east of Inchkeith in 12 fathoms.

F. carbasea (Ell. and Sol.).
Leith (*Colds.*); Leith Shore (*Mr Parsons*);

Firth of Forth (*M'B.*); oyster beds, Firth of Forth (*Grant*); Newhaven, on the fishermen's nets (*Dr Landsborough*).

This species is common. We have obtained it frequently from the dredges at Newhaven pier, and took it in 7 fathoms, off Fidra, last summer.

MEMBRANIPORIDÆ.

Membranipora catenularia (James.).
: Firth of Forth (*Ed. Mus.*).
This is the *Hippothoa catenularia* of Johnston and other authors.

M. pilosa (Linn.).
: On *Laminaria*, etc. (*M'B.*); Firth of Forth (*Ed. Mus.*); off Fidra, 7 fathoms (*nob.*).
This species is abundant, usually on the larger Algæ, in most parts of the Firth; it is often found cast ashore after storms.

M. membranacea (Linn.).
: Also extremely abundant on Algæ, especially *Laminaria* fronds; on stones at low water (*M'B.*). Off Fidra, 7 fathoms (*nob.*).

M. lineata (Linn.).
: Firth of Forth (*C. W. P.*); Firth of Forth (*M'B.*).

M. flemingi (Busk).
: Newhaven (*C. W. P.*).

CRIBRILINIDÆ.

Cribrilina radiata (Moll).
: On shells and stones (*M'B.*).
The *Lepralia innominata* of Johnston, etc.

C. punctata (Hassall).
: Firth of Forth (*C. W. P.*); Firth of Forth (*Ed. Mus.*).
The *Lepralia punctata* of Johnston and Busk.

Membraniporella nitida (Johnst.).
>Firth of Forth (*C. W. P.*) ; on shells, rocks, etc. (*M'B.*).
>The *Lepralia nitida* of Johnston and Busk.

MICROPORELLIDÆ.

Microporella ciliata (Pall.).
>Firth of Forth (*C. W. P.*); on rocks and *Fuci* (*M'B.*).
>The *Lepralia ciliata* of Johnston and Busk.

M. malusii (Audouin).
>Newhaven (*C. W. P.*); on shells, etc. (*M'B.*).
>The *Lepralia biforis* of Johnston.

MYRIOZOIDÆ.

Schizoporella linearis (Hassall).
>Newhaven (*C. W. P.*); Firth of Forth (*Ed. Mus.*).
>The *Lepralia linearis* of Johnston and Busk.

S. auriculata (Hassall).
>Firth of Forth (*C. W. P.*).
>The *Lepralia auriculata* of Johnston and Busk.

S. hyalina (Linn.).
>On Algæ, stones, and corallines (*M'B.*).

Hippothoa divaricata (Lamk.).
>We dredged this species in the Firth last summer.

ESCHARIDÆ.

Porella compressa (Sowerby).
>Fifeshire coast, rare (*J. G.*).
>The *Cellepora cervicornis* of Johnston, Busk, and other authors.

Smittia trispinosa (Johnst.).
>Firth of Forth (*C. W. P.*).
>The *Lepralia trispinosa* of Johnston.

Mucronella peachi (Johnst.).
> Firth of Forth (*C. W. P.*); on rocks and *Fuci* (*M'B.*); Firth of Forth (*Ed. Mus.*).
> The *Lepralia peachii* and *L. immersa* of Johnston.

M. ventricosa (Hassall).
> Firth of Forth (*C. W. P.*); Firth of Forth (*Ed. Mus.*).
> The *Lepralia ventricosa* of Johnston and Busk.

M. coccinea (Abildg.).
> On stones, etc., Firth of Forth (*M'B.*).

M. variolosa (Johnst.).
> Firth of Forth (*C. W. P.*); on bivalve shells and stones (*M'B.*); Firth of Forth (*Ed. Mus.*).
> The *Lepralia variolosa* of Johnston and Busk.

Palmicellaria skenii (Ell. and Sol.).
> Firth of Forth (*C. W. P.*).
> The *Cellepora skenei* of Johnston, Busk, and others.

CELLEPORIDÆ.

Cellepora pumicosa (Linn.).
> Firth of Forth (*Ed. Mus.*); on corallines, stones, and *Fuci* (*M'B.*).
> We dredged this species last summer.

C. ramulosa (Linn.).
> Firth of Forth (*Ed. Mus.*).

CYCLOSTOMATA.

CRISIIDÆ.

Crisia cornuta (Linn.).
> We have taken this species, the *Crisidia cornuta* of Johnston's "British Zoophytes," in the Firth.

Crisia eburnea (Linn.).
    Firth of Forth (*M'B.*).
C. denticulata (Lamk.).
    Firth of Forth (*M'B.*).
    Not uncommon in the Firth. We dredged it last summer, south-west of Inchkeith, in 5 fathoms; off Aberdour, in 5 fathoms; and off Aberlady Bay, in 9 fathoms.

TUBULIPORIDÆ.

Stomatopora dilatans (Johnst.).
    Firth of Forth (*C. W. P.*).
S. incurvata (Hincks).
    Outside the Isle of May (*C. W. P.*).
Idmonea serpens (Linn.).
    On Zoophytes, Firth of Forth (*M'B.*).
    We dredged this species to the north-west of the Isle of May in 26 fathoms.
Diastopora patina (Lamk.).
    On the fishermen's lines, Newhaven (*C. W. P.*); south-west of Inchkeith, 5 fathoms (*nob.*); on shells, Zoophytes, rocks, and Algæ, Firth of Forth (*M'B.*).

LICHENOPORIDÆ.

Lichenopora hispida (Flem.).
    On Algæ, Firth of Forth (*M'B.*).
L. radiata (Audouin).
    Firth of Forth (*C. W. P.*).

CTENOSTOMATA.

ALCYONIDIIDÆ.

Alcyonidium gelatinosum (Linn.).
    Bass Rock, 24 fathoms (*K.*); Firth of Forth (*M'B.*); Firth of Forth (*Ed. Mus.*); off the Isle of May (*nob.*).
    This species is common in the Firth, in a few fathoms, attached usually to dead

shells. We have also taken it at low water on Cramond Island.

Alcyonidium hirsutum (Flem.).

On *Fucus serratus*, Firth of Forth (*M'B.*); Entrance to Firth of Forth, 30 fathoms (*K.*).

We dredged this species last summer off Aberlady Bay in 9 fathoms.

A. mytili (Dalyell).

We have taken this species in the Firth.

A. parasiticum (Flem.).

Firth of Forth (*M'B.*); Firth of Forth (*Ed. Mus.*); entrance to Firth of Forth, 30 fathoms; Firth of Forth, 24 fathoms; Bass Rock, 24 fathoms (*K.*).

We have obtained this species north-east of Inchkeith, 12 fathoms; west of Inchkeith, 5 fathoms; off Aberlady Bay, 9 fathoms; Kirkcaldy Bay, 9 fathoms; off Aberdour, 5 fathoms; and frequently at Newhaven. It is common the Firth.

A. polyoum (Hassall).

Firth of Forth (*M'B.*).

The *Sarcochitum polyoum* of Johnston and others.

FLUSTRELLIDÆ.

Flustrella hispida (Fabr.).

On *Fucus serratus*, Firth of Forth (*M'B.*). Common on *Fucus* at low water mark, Aberdour.

VESICULARIIDÆ.

Vesicularia spinosa (Linn.).

Firth of Forth (*M'B.*); Leith shore (*D. Landsb., jun.*).

We have dredged this species in abundance west of Inchkeith in 5 fathoms, and have obtained it frequently from the dredges at Newhaven. It is often found on the shore after storms, deprived of the polypites.

Amathia lendigera (Linn.).
Firth of Forth (*D. Landsb., jun.*); Firth of Forth (*M'B.*); Firth of Forth (*Ed. Mus.*).
The *Serialaria lendigera* of Johnston.

Bowerbankia imbricata (Adams).
This species is common in the Firth, usually in tangled masses among other Polyzoa or Zoophytes. We have dredged it frequently.

B. pustulosa (Ell. and Sol.).
Leith Shore, rare (*D. Landsb., jun.*).

Avenella fusca (Dalyell).
Newhaven, among rejectamenta of the oyster dredges, on corallines, etc. (Sir C. Wyville Thomson, *Ann. N. H.*, 1852).

VALKERIIDÆ.

Valkeria uva (Linn.).
Leith Shore (*R. J.*); Firth of Forth (*M'B.*).
We have dredged this species frequently in the Firth, and have also obtained it between tide marks at Newhaven and elsewhere.

ENTOPROCTA.

PEDICELLINIDÆ.

Pedicellina cernua (Pallas).
Firth of Forth (*M'B.*).
This species, the *Pedicellina echinata* of Johnston and others, is not uncommon in the Firth. We have taken both the echinated and the smooth (*P. belgica*, Gosse) variety. It occurs in profusion on the tests of *Stycla grossularia* under large stones at Newhaven.

CRUSTACEA.

The groups which the following list overtakes are the Cirripedia, Amphipoda, Isopoda, Cumacea, Stomapoda, and Decapoda. A very full list of the Ostracoda of the east coast of Scotland, collected from the Aberdeenshire coast, Montrose, the Firth of Forth, etc., will be found in a paper by Professor G. S. Brady and Mr David Robertson, " On the Distribution of the British Ostracoda."* We shall not attempt at present to treat either this group or the Copepoda.

In the arrangement and nomenclature of the Amphipoda and Isopoda, we have followed Bate and Westwood's " British Sessile-eyed Crustacea," a work from which we have derived the greatest assistance.

We have used the lists, prepared by Metzger, of the Crustacea obtained by the German Exploring Expedition of 1872. Many of the species in that report were not previously recorded from the Forth.

CIRRIPEDIA.

I. SUCTORIA—

PELTOGASTRIDÆ.

Peltogaster paguri (Rathke).

Firth of Forth, Joppa (J. Anderson, M.D., *Proc. Roy. Phys. Soc.*, vol. ii.).

We have found this species on *Pagurus bernhardus* not unfrequently.

P. carcini (Ander.).

Firth of Forth, Joppa (J. Anderson, M.D., *Proc. Roy. Phys. Soc.*, vol. ii.).

Sacculina carcini (Thomps.).

Trinity (*Com. Mar. Zool.*); Firth of Forth (*J. Anderson, M.D.*); Firth of Forth (*Ed. Mus.*).

Often attached to the abdomen of *Carcinus mænas.*

S. triangularis (Ander.).

Firth of Forth (J. Anderson, M.D., *Proc. Roy. Phys. Soc.*, vol. ii.).

* Ann. and Mag. Nat. Hist., Ser. 4, vol. ix.

II. THORACICA—
LEPADIDÆ.
Lepas anatifera (Linn.).
   Attached to floating timber, Firth of Forth (*Ed. Mus.*).
   We have found this species cast ashore at North Berwick.
Conchoderma virgata (Spengl.).
   On floating timber, Firth of Forth (*Ed. Mus.*).
C. aurita (Linn.).
   On floating timber, Firth of Forth (*Ed. Mus.*).

BALANIDÆ.
Balanus balanoides (Linn.).
   Firth of Forth (*Ed. Mus.*).
   This is an exceedingly abundant species between tide marks, and we have dredged it very frequently in pretty deep water.
B. porcatus (E. da Costa).
   Not uncommon attached to stones, etc.
B. crenatus (Brug.).
   Portobello (*Ed. Mus.*).
B. tintinnabulum (Linn.).
   Leith Harbour, foreign importation (*Ed. Mus.*).
   This species is an inhabitant of the warmer seas, and its occurrence in the Forth is an accidental circumstance.

EDRIOPHTHALMATA.
AMPHIPODA.
I. SALTATORIA—
ORCHESTIIDÆ.
Talitrus locusta (Linn.).
   Firth of Forth (*Ed. Mus.*).
   Very abundant about high tide mark among stones, sea-weed, etc.

II. NATATORIA—

GAMMARIDÆ.

Callisoma kröyeri (Bruzel).
    Mouth of the Firth, 30 fathoms (*Metzger*).
Ampelisca macrocephala (Lilljeborg).
    Firth of Forth, 24 fathoms (*Metzger*).
A. tenuicornis (Lilljeborg).
    Bass Rock, 24 fathoms; off St Abb's Head, 40 fathoms (*Metzger*).
Protomedeia fasciata (Kröyer).
    St Abb's Head, 40 fathoms (*Metzger*).
Melita obtusata (Mont.).
    Bass Rock, 24 fathoms; off St Abb's Head, 40 fathoms (*Metzger*).
Gammarus locusta (Linn.).
    Very abundant in rock pools, and generally between tide marks.

COROPHIIDÆ.

Podocerus capillatus (Rathke).
    We have dredged this species in 5 fathoms off Inchkeith.
Cerapus difformis (Milne-Edwards).
    Bass Rock (*Metzger*).
Corophium longicorne (Latr.).
    Dunbar (*Mr David Robertson*).

III. ABERRANTIA—

CAPRELLIDÆ.

Protella phasma (Mont.).
    Isle of May (*Brit. Mus.*); Firth of Forth (*H. D. S. G.*).
Caprella linearis (Linn.).
    Plentiful in the upper laminarian zone, and we have also dredged it in a few fathoms.
C. lobata (Müll.).
    Firth of Forth (*Brit. Mus.*).
    The *C. laevis* of Goodsir.

Caprella acanthifera (Leach).
>Firth of Forth (*Bell Collection, Oxford*, Rev. J. Gordon).

C. tuberculata (Guerin).
>Firth of Forth (*Brit. Mus.*).

C. typica (Kröyer).
>Firth of Forth (*Bell Collection, Oxford*).

ISOPODA.

NORMALIA—

BOPYRIDÆ.

Phryxus abdominalis (Kröyer).
>Off St Abb's Head, 40 fathoms (*Metzger*).

P. paguri (Rathke).
>Firth of Forth (J. Anderson, M.D., *Proc. Roy. Phys. Soc.*, vol. ii.).

Cryptothiria balani (S. Bate).
>The female of this species was described by Mr H. Goodsir, from the Forth, as the male of *Balanus balanoides*.

ASELLIDÆ.

Munna kröyeri (H. Goods.).
>Firth of Forth (*Brit. Mus.*).
>This species was discovered in the Forth by Mr H. Goodsir.

Limnoria lignorum (Rathke).
>We obtained it at Elie.

ARCTURIDÆ.

Arcturus longicornis (Sowerby).
>Firth of Forth (*Brit. Mus.*); Firth of Forth (*Ed. Mus.*); Bass Rock, 24 fathoms (*Metzger*).
>The specimens from which this species was described and figured were obtained by Mr Simmons, near Inchkeith. We have frequently dredged this species in many parts of the Firth.

Arcturus intermedius (H Goods.).
>Firth of Forth (*Brit. Mus.*).
>This species was first found, opposite Anstruther, by Mr H. Goodsir, and was described by him as *Leachia intermedia*.

A. gracilis (H. Goods.).
>Firth of Forth (*Brit. Mus.*).
>Obtained by Mr H. Goodsir off Anstruther, and described by him as *Leachia gracilis*.

Idotea tricuspidata (Desmarest).
>We have obtained this species in shallow water at various localities in the lower basin of the Forth.

CUMACEA.

DIASTYLIDÆ.

Cuma edwardsii (H. Goods.).
>Firth of Forth (*H. D. S. G.*).
>We obtained this species in Largo Bay.

C. scorpioides (Mont.).
>Firth of Forth (*H. D. S. G.*).

C. trispinosa (H. Goods.).
>Firth of Forth (*H. D. S. G.*).

Alauna rostrata (H. Goods.).
>Firth of Forth (*H. D. S. G.*).

Bodotria arenosa (H. Goods.).
>Firth of Forth (*H. D. S. G.*).

PODOPHTHALMATA.

STOMAPODA.

MYSIDÆ.

Mysis flexuosa (O. F. Müll.).
>Rock pools, Seafield (*M'B.*); Firth of Forth (Leach, in *Brit. Mus.*).

Cynthia flemingi (H. Goods.).
>Firth of Forth (*H. D. S. G.*); Firth of Forth (*Brit. Mus.*).

Themisto longispinosa (H. Goods.).
    Firth of Forth (*H. D. S. G.*).
T. brevispinosa (H. Goods.).
    Firth of Forth (*H. D. S. G.*).

DECAPODA.

I. MACRURA—

ASTACIDÆ.

Homarus gammarus (Linn.).
    Firth of Forth at low water, many places (*Howd.*).
    Caught in considerable numbers for the markets on all the rocky shores near the mouth of the estuary.

Nephrops norvegicus (Linn.).
    Firth of Forth (Leach, in *Brit. Mus.*); Largo, Leith, etc. (*Howd.*).
    Very abundant near the mouth of the Firth, where immense numbers are got by the trawlers for the markets. We have obtained it alive near Aberdour.

CRANGONIDÆ.

Crangon vulgaris (Fabr.).
    On sandy beaches, Seafield (*Howd.*).
    Common on all the sandy shores.

C. allmani (Kinahan).
    Bass Rock, 24 fathoms (*Metzger*).

C. nanus (Kröyer).
    Bass Rock, 24 fathoms (*Metzger*).

PALÆMONIDÆ.

Hippolyte spinus (Sowerby).
    Newhaven (*Leach*); Firth of Forth (Dr Neill, in *Brit. Mus.*).
    This species is rather common in the laminarian and littoral zones.

H. varians (Leach).
    Firth of Forth, in pools (*Howd.*).

Hippolyte securifrons (Norman).
> Off St Abb's Head, 40 fathoms (*Metzger*).

Pandalus annulicornis (Leach).
> Black Rocks, Leith; Seafield (*M'B.*); Bass Rock, 24 fathoms; St Abb's Head, 4 fathoms (*Metzger*).
> Common and generally distributed. We have dredged it off Inchkeith, 5 fathoms, and in Aberlady Bay, and in many other localities.

Palæmon squilla (Linn.).
> Frequent in rock pools near the mouth of the Firth.

II. ANOMURA—

LITHODIDÆ.

Lithodes maia (Linn.).
> Dunbar (*Ed. Mus.*); Firth of Forth (*Howd.*); young, from stomach of cod (Dr Neill, in *Brit. Mus.*).
> This species is not uncommon near the mouth of the Firth. It is often obtained by fishermen near the Isle of May.

PAGURIDÆ.

Pagurus bernhardus (Linn.).
> Firth of Forth (*Ed. Mus.*); Firth of Forth common (*Howd.*).
> Extremely abundant in the Forth, especially on the oyster bank, where it is found in numbers in every dredgeful. It attains a large size, and when adult usually inhabits the shell of *Buccinum undatum*.

P. ulidianus (Thompson).
> Firth of Forth (*Howd.*).

P. hyndmanni (Thompson).
> Firth of Forth (*Ed. Mus.*); Musselburgh and Prestonpans (*Howd.*).

Pagurus lævis (Thompson).
> Firth of Forth (*Howd.*).

P. forbesii (Bell).
> Firth of Forth (*Howd.*).

PORCELLANIDÆ.

Porcellana platycheles (Penn.).
> Firth of Forth (*Ed. Mus.*); Crail and Fifeness at low water (*Howd.*).
> We have found this littoral species at Elie, and on the shore near North Berwick.

P. longicornis (Penn.).
> Upper part of the Firth (*Howd.*); Bass Rock, 24 fathoms (*Metzger*).
> We have dredged it off the Isle of May in 8 fathoms, near Elie, and near Inchkeith.

GALATHEIDÆ.

Galathea squamifera (Mont.).
> Common under stones (*M'B.*).
> This is a very common littoral species, but extends into the laminarian zone. We have dredged it on the oyster bank.
> We found this species frequently at low water on the shore near Elie.

G. andrewsii (Kinahan).
> Firth of Forth (J. Anderson, M.D., in *Proc. Roy. Phy. Soc.*, vol. ii.).

G. strigosa (Penn.).
> Firth of Forth (*Ed. Mus.*); off the Bass Rock (*Howd.*).
> Not uncommon near the mouth of the Firth. Our friend, Mr Robert Gray, F.R.S.E., informs us that it is plentiful near Dunbar.

G. nexa (Embleton).
> Off Portseaton (*Howd.*).

Munida bamffica (Penn.).
> The *M. rondeletii* of Bell and other authors.

Mr Robert Gray found it not uncommon at Dunbar.

III. BRACHYURA—

LEPTOPODIADÆ.

Stenorhynchus rostratus (Linn.).

Prestonpans (*Ed. Mus.*); Firth of Forth on mud and sand, generally distributed (*Howd.*).

This species, the *S. phalangium* of authors, is not uncommon. We have dredged it on the oyster bank at 5 fathoms, in Aberlady Bay, near Elie, and elsewhere.

MAIADÆ.

Inachus dorsettensis (Penn.).

Deep-sea lines (*Howd.*).

Hyas araneus (Linn.).

Very abundant. We have often taken it between tide marks at Newhaven, Aberdour, and other places, and have dredged it off Longniddry in 14 fathoms, and in Aberlady Bay and on the oyster bank in 5 fathoms. It occurs in almost every dredgeful in the lower reaches of the Firth.

H. coarctatus (Leach).

Largo Bay, Inchkeith, etc. (*M'B.*); Firth of Forth (Leach, in *Brit. Mus.*).

This species was first discovered in the Firth of Forth by Dr Leach. It is fairly common but less plentiful than *H. araneus*. We have obtained it south-west of Inchkeith, 5 fathoms, in Aberlady Bay, 5 fathoms, and in many other localities.

PARTHENOPODÆ.

Eurynome aspera (Penn.).

Prestonpans and Portseaton (*Howd.*).

This deep-water form is rare in the Firth.

CANCERIDÆ.

Cancer pagurus (Linn.).
  Firth of Forth (*Ed. Mus.*).
  Common in the laminarian and littoral zones.

PORTUNIDÆ.

Carcinus mænas (Linn.).
  Firth of Forth (*Ed. Mus.*).
  Very abundant between tide marks, and in the laminarian zone, but not often found in deep water.

Portumnus variegatus (Leach).
  Prestonpans and Portseaton (*Howd.*).
  We have taken this species at Portobello.

P. puber (Linn.).
  We obtained one specimen on the deep-sea lines, from the mouth of the Forth.

P. depurator (Linn.).
  Firth of Forth (*M'B.*).
  A very common species on the oyster banks, and often cast ashore.

P. marmoreus (Leach).
  Portobello and Musselburgh beaches (*Howd.*).

P. holsatus (Fabr.).
  Newhaven (*Brit. Mus.*).
  This is the *P. lividus* of Leach, who found one amongst a number of specimens of *P. depurator* at Newhaven.

P. pusillus (Leach,.
  Off Prestonpans (*Howd.*); Firth of Forth (*Brit. Mus.*).
  Fairly common in the Forth. We have very frequently dredged it near Inchkeith, etc.

PINNOTHERIDÆ.

Pinnotheres pisum (Penn.).
Firth of Forth (*M'B.*).
Not uncommon; we generally found it in the pallial chamber of *Modiola modiolus*. We have dredged it off Longniddry in 14 fathoms, and elsewhere.

LEUCOSIADÆ.

Ebalia cranchii (Leach).
Firth of Forth, rare (*H. D. S. G.*).

CORYSTIDÆ.

Atelecyclus septemdentatus (Mont.).
Firth of Forth (*Ed. Mus.*); Firth of Forth, rare (*H. D. S. G.*); Portobello beach (*M'B.*). The *A. heterodon* of Leach. This species seems to be a very favourite food of the cod. Dr J. A. Smith has recorded it from the stomach,* and we have found it in the same situation.

Corystes cassivelaunus (Penn.).
Firth of Forth (*Ed. Mus.*); Off Inchkeith (*M'B.*); Bass Rock, 24 fathoms (*Metzger*); Aberlady Bay (*Com. Mar. Zool.*); Newhaven (*C. W. P.*).
Generally distributed; we have dredged it in Aberlady Bay, 5 fathoms, and in Kirkcaldy Bay, 9 fathoms.

TUNICATA.

There is practically no literature on the Tunicata of the Firth of Forth. Comparatively few naturalists have worked at this interesting but obscure group, while amateurs, to whom we are so often indebted for valuable contributions to local faunas, invariably fail us when we come to the Ascidians.

The number of species recorded from the Firth of Forth is

* Proc. Phy. Soc., vol. iii., p. 214.

very small, but it must be further reduced, because of the impossibility of determining with certainty the species referred to by many of the older writers, on account of the confusion which has existed between some of the allied forms, and the tangled mass of synonymy in which other species have become hopelessly involved. Accordingly, with the exception of Alder's two species of *Parascidia* from the Isle of May, and of *Pelonaia corrugata* (F. and G.), which was first discovered by Professor Goodsir, in deep water, off Anstruther, in 1841, and which was dredged near the Bass Rock in 24 fathoms by the German North Sea Expedition of 1872, we have given in the following list only those species which we have ourselves collected in the Firth of Forth.

ASCIDIÆ SIMPLICES.

MOLGULIDÆ.

Molgula citrina (Alder and Hancock).

This little species was first described by Alder in his "Catalogue of the Marine Mollusca of Northumberland and Durham,"[*] and, so far as we are aware, it has not been mentioned since.

We have come upon it several times during the last two years, adhering to the under surfaces of large stones, about low water mark, between the Chain Pier and Granton Harbour.

Eugyra glutinans (Möller).

This species has a most extensive synonymy, and is usually known as *Molgula* (or *Eugyra*) *arenosa* (Ald. and Han.). Lately, however, Traustedt[†] has declared that it is identical with the species described in 1842 by Möller[‡] as *Cynthia glutinans*.

---

[*] Trans. Tyneside Nat. Field Club, vol. i., p. 199 (1850).

[†] Oversigt o. d. f. Danmark, etc., Asc. Simp. (Vid. Medd. nat. For. Kbhvn, 1879-80).

[‡] Index. Moll. Groenl., 1842, p. 21.

This species occurs on sandy bottoms, and appears to be gregarious, as a considerable number of specimens are usually obtained together.

We dredged it plentifully last summer off Kirkcaldy Bay in 9 fathoms.

CYNTHIIDÆ.

Styela grossularia (van Beneden).

This species was formerly considered as the young of *Styela rustica* (O. F. Müller), and is probably the species referred to under the name of *Cynthia rustica* by M'Bain.

It is very common in the Firth of Forth, from the shore out to deep water. It occurs in profusion between tide marks at Newhaven, Wardie, Aberdour, etc., covering the under surfaces of stones, and often in such abundance as to form large masses of individuals adhering by their tests. The tests are frequently so closely united as to appear like a common investing mass in which a colony of individuals is imbedded. It is merely an aggregation, however, and gemmation seems never to take place. On the oyster bank this species is found on dead shells, etc. Here, though still common, the individuals are not crowded together, and each is enabled to preserve its characteristic blister-like shape, and to develop the spreading margin, which is rarely seen in specimens from Newhaven.

We have dredged this species in other parts of the Forth, such as—east of Inchkeith, 7 fathoms; off Kirkcaldy Bay, 9 fathoms; Aberlady Bay, 9 fathoms; off Longniddry, 14 fathoms; off Aberdour,

5 fathoms; and have collected it between tide marks at several points on both shores of the Firth.

Pelonaia corrugata (Forb. and Goods.).

This species was dredged by Professor Goodsir in deep water off Anstruther, and was described first in Jameson's *Edinburgh New Philosophical Journal* for 1841 (vol. xxxi., p. 29). The only occasion, so far as we are aware, on which it has since been taken in the Firth of Forth was in 1872, when it was dredged by the German North Sea Expedition in 24 fathoms, off the Bass Rock.* Dr M'Intosh informs us that he has obtained this species several times at St Andrews. It has also been found in Berwick Bay and down the Northumberland coast.

ASCIDIIDÆ.

Ascidia virginea (O. F. Müller).

This species, which is synonymous with *Ascidia sordida* (Ald. and Han.), is very common in a few fathoms of water and upwards in all parts of the Firth. On the oyster banks it occurs in profusion adhering to dead shells, Algæ, Zoophytes, etc. The specimens are of large size, and often united in clumps. We have also dredged it plentifully in other parts of the Firth, viz., north-east of Inchkeith, 12 fathoms; Aberlady Bay, 9 fathoms; off the Isle of May, 8 fathoms; and off Fidra, 7 fathoms.

Young specimens of this species are often obtained adhering in clusters to masses of *Gemellaria loricata*, and in this condition are frequently found on the beach after

* Kupffer—Jahresb., VII. Tunicata, p. 227.

storms; they are probably what are referred to by Dr M'Bain as *Ascidia prunum*. These young specimens are perfectly transparent, and have a beautiful crystalline appearance. They are excellent objects in which to study the circulation in the living animal.

Ascidia depressa (Alder).

We obtained several specimens of this characteristic species, some years ago, at low water mark near Elie.

A. scabra (O. F. Müller).

This species we also obtained on the same occasion, near Elie, in considerable quantity.

Ciona intestinalis (Linnæus).

This beautiful species is fairly common on the oyster bank, and on muddy bottoms elsewhere throughout the Firth; it attains a large size. We have dredged it in various parts of the Firth, and have also taken it at low water mark at Elie and Aberdour, generally adhering to the roots of *Laminaria*.

ASCIDIÆ COMPOSITÆ.

BOTRYLLIDÆ.

Botryllus schlosseri (Pallas).

This species is common at low water mark on the shores of the Firth. It is usually found on the under surface of large stones, or encrusting the roots and stems of *Laminaria*, *Fucus*, etc.

We have taken it at Elie, North Berwick, Aberdour, and other localities.

Dr M'Bain records this species in his list.

B. polycyclus (Savigny).

This species appears to be rarer than the last.

Dr M'Bain mentions it in his list, and we dredged some large specimens in 10 to 20 fathoms, east of Inchkeith, last summer.

Botryllus calendula? (Giard).

We refer provisionally to this species some specimens of a rather delicate *Botryllus* which we dredged last summer in 14 fathoms, off Longniddry. The specimens were encrusting large individuals of *Ascidia virginea.*

Botryllus calendula was described by Giard * from Roscoff, on the coast of Brittany, where it is stated to be very rare.

Botrylloides rubrum (Milne-Edwards).

The specimens which we have found in the Firth of Forth have belonged to the yellow varieties† of the species, and not to the typical red form. They were from Elie, and were taken at low water mark at spring tides.

B. radiata? (Alder and Hancock).

A few specimens which we found along with the last species may be referred provisionally to *Botrylloides radiata.* It is possible that this species may be merely a variety of *B. rotifera* (M. Edw.).

B. albicans (Milne-Edwards).

We have obtained this species at low water mark near Newhaven.

POLYCLINIDÆ.

Aplidium caliculatum (Savigny).

We obtained a large specimen in an excellent state of preservation from the stomach of a cod last summer.

This species, having a long filiform post-abdomen, a channelled stomach, and a

---

* Recherches sur les Synascidies (Arch. Zool. expér., t. 1, p. 623, 1872).
† See Giard, *loc. cit.*, p. 632.

six-rayed branchial aperture, falls into the sub-genus *Amaroucium* of Giard's system of nomenclature.

Aplidium proliferum (Milne-Edwards).

This species also belongs to the sub-genus *Amaroucium*. It is not uncommon at low water mark and in the upper laminarian zone at Elie.

In the *Edinburgh New Philosophical Journal* for 1838 (vol. xxvi., p. 155), Sir John Dalyell gives an account, under the name of *Aplidium verrucosum*, of a compound Ascidian which was dredged to the south-east of Inchkeith, and brought to him by the fishermen.

Parascidia forbesi (Alder).

This species, the *Sidnyum turbinatum* of Forbes, and *Circinalium concrescens* of Giard, was obtained from the Isle of May.

P. flemingi (Alder).

Also from the Isle of May on rocks (*Flem.*).

DIPLOSOMIDÆ.

Pseudodidemnum gelatinosum (Milne-Edwards).

The *Didemnum gelatinosum* of Milne-Edwards and other authors.

We have taken small colonies of this species near low water mark at Wardie.

We hope in time to add largely to the above list of Tunicata. There are many species known to inhabit the North Sea which we may reasonably expect to find in the Firth of Forth, especially towards the mouth. *Clavelina lepadiformis* is recorded from St Andrews by Dr M'Intosh, and from the Northumberland coast by Mr Hancock; it will doubtless be found within our area also. Other species, such as *Botrylloides leachi* (St Andrews, M'Intosh; and Northumberland, Alder), *Polyclinum aurantium* (Cullercoats, Hancock), and *Polycarpa tuberosa* (Aberdeen, Macgillivray; and Cullercoats, Alder), will, we expect, be also obtained when the Firth of Forth, and especially its outer part, has been more thoroughly worked.

PORIFERA.

The nomenclature of the Fibrous Sponges adopted in this article is that of Professor Oscar Schmidt, as given in his "Grundzüge einer Spongien-Fauna des Atlantischen Gebietes," Leipzig, 1870. A number of the specimens we have found in the Forth are not yet identified.

CALCAREA.

Grantia compressa (Fabr.).
Not uncommon in the littoral zone. We have obtained it at Elie, Wardie, and in other parts of the shore on both sides of the estuary.

G. ciliata (Fabr.).
Frequent on roots of *Laminaria*, in the lower littoral zone, at North Berwick, etc.

Leuconia nivea (Grant).
Under surface of sheltered rocks, Prestonpans Bay (*Flem.*).
We have obtained this species at Aberdour, and near Elie.

FIBROSA.

CHALINIDÆ.

Chalinula oculata (Pallas).
We have found this species on the pier at Newhaven. It is the *Chalina oculata* of Bowerbank.

Isodictya infundibuliformis (Bowerb.).
Firth of Forth (*Ed. Mus.*).

I. simulans (Johnst.).
We obtained this species at Elie.

I. palmata (Johnst.).
We have dredged this species in the Firth, and have also found it on Newhaven pier.

RENIERIDÆ.

Amorphina panicea (Pallas).
: Firth of Forth (*Ed. Mus.*).
  The *Halichondria panicea* of Bowerbank.
  It is very abundant, encrusting rocks about low water mark.

A. coalita (O. F. Müll.).
: Firth of Forth, very common (*Flem.*); Firth of Forth (*Ed. Mus.*).
  The *Halichondria coalita* of Bowerbank. We have dredged it in the Forth on several occasions.

A. paciscens (O. Schmidt).
: This species was first described from a specimen found on the beach at Portobello by the German Expedition of 1872.

SUBERITIDÆ.

Suberites domuncula (Nardo).
: Firth of Forth (*Flem.*); beach, Portobello (*O. Schmidt*).
  The *Hymeniacidon suberea* of Bowerbank.

Vioa celata (Grant).
: The *Hymeniacidon celata* of Bowerbank.
  It is very common in dead shells, on the oyster bank, and elsewhere.

CTENOPHORA.

I. SACCATÆ—

CYDIPPIDÆ.

Pleurobrachia pileus (Eschsch.).
: Firth of Forth (*F. E. S.*).
  We have obtained this species in abundance in the late summer and autumn months in Granton Harbour, Leith Roads, Elie, and elsewhere. It usually occurs in swarms.

II. EURYSTOMÆ—

BEROIDÆ.

Idyia ovata (Less.).
Firth of Forth (*F. E. S.*).
We have obtained this form in the autumn, off Elie. It is the *Beroe ovata* of Eschscholtz.

ACALEPHA.

Aurelia aurita (Linn.).
Very common in the autumn months; often stranded on the beach.

Cyanea capillata (Eschsch.).
Common in all the lower reaches of the Firth, in the autumn.

ZOANTHARIA.

I. ACTINARIA—

SAGARTIADÆ.

Actinoloba dianthus (Ellis).
Firth of Forth, on rocks uncovered only at very low tides (*T. S. W.*); Firth of Forth (*Colds.*); Firth of Forth (*M'B.*).

Sagartia troglodytes (Johnst.).
Firth of Forth (*T. S. W.*); Firth of Forth (*M'B.*).
This species is commonly found on the shore at very low water. We have obtained it at Elie, Aberdour, and elsewhere.
The variety *S. prasina* of Gosse, found by Dr Strethill Wright in the Forth, is an inhabitant of deeper water. We have dredged a considerable number of specimens of it in the neighbourhood of Inchcolm in 10 to 18 fathoms. It is characterised by having the disc and tentacles of a bright green colour.

Sagartia ornata (Holdsworth).
Firth of Forth (*M'B.*).

ACTINIADÆ.

Actinia mesembryanthemum (Ellis and Sol.).
Leith shore (*R. J.*); Firth of Forth (*M'B.*). This is a very abundant species in the lower littoral zone, on exposed rocks and in rocky ledges of pools left by the retiring tide. A well-grown specimen was taken from the Forth in 1828, by Sir John Graham Dalyell, who kept it for twenty years. It then passed successively into the hands of Professor Fleming and of Dr M'Bain. It is now in the possession of Mr Sadler, curator of the Royal Botanic Garden, Edinburgh, and continues in vigorous health.

The variety *A. chiorocca* was found near the Bass Rock by the German North Sea Expedition.

BUNODIDÆ.

Tealia crassicornis (O. F. Müll.).
Firth of Forth (*M'B.*); Firth of Forth, 30 fathoms (*F. E. S.*).
Common in the littoral and upper laminarian zones, and extending into deeper water. We have dredged it on the oyster banks and elsewhere. It sometimes attains a very large size—a specimen which we found near Aberdour measuring nearly a foot in diameter.

The *Actinia gemmacea*, Leith shore (*R. J.*), was probably this species.

ILYANTHIDÆ.

Halcampa crysanthellum (Peach).
Bass Rock, 24 fathoms (*F. E. S.*).

Halcampa fultoni (T. S. W.).
Granton pier (*T. S. W.*).

The following MEDUSOID GONOPHORES have been noticed in the Firth of Forth:

Oceania pileata (Forskal).
Firth of Forth, surface (*F. E. S.**).
Lizzia sp.
Mouth of Forth, surface (*F. E. S.*).
Eucope lucifera (Forbes).
Firth of Forth (*F. E. S.*).
Phalidium viridicans (Leuckhart).
Firth of Forth (*F. E. S.*).
Tima bairdi (Johnst.).
Burntisland Harbour, winter (*E. F.*).
Bougainvillea brittanica (Forbes).
Entrance to Firth (*E. F.*).
Taken in surface-net cast of Inchkeith last summer (*nob.*).
Goodsirea mirabilis (T. S. W.).
Near Queensferry (T. S. W., in *Proc. Roy. Phys. Soc.*, vol. iii.).
Stomobrachium octocostatum (Forbes).
Near Queensferry and Granton (T. S. W., in *Proc. Roy. Phy. Soc.*, vol. iii.).
Acanthobrachia inconspicua (T. S. W.).
Granton Harbour (T. S. W., in *Proc. Roy. Phy. Soc.*, vol. iii.).

VERMES.

A very complete list of the Vermes of the eastern coast of Scotland will be found in Dr M'Intosh's "Marine Fauna of St Andrews." We have worked little at this interesting but difficult group, and the following is a very imperfect list of the forms to be found in the Forth. In its preparation we have derived much assistance from Dr M'Intosh's work, and have followed his arrangement of the group. We have

---

* II. Jahresb. d. Komm. z. Untersuch. d. deutsch. Meere, III., Coelenterata.

used the lists prepared by Professor Möbius, of the worms obtained in the German Expedition of 1872, and for the list of species from the Forth in the British Museum, we are indebted to Johnston's Catalogue of non-Parasitical Worms.

CHÆTOGNATHA.

Sagitta bipunctata (Quoy and Gaimard).
Off St Abb's Head, 3 fathoms (*Möb.*).

GEPHYREA.

SIPUNCULIDÆ.

Phascolosoma strombi (Mont.).
Bass Rock (*Möb.*).
P. procerum (Möb.).
Bass Rock, 24 fathoms (*Möb.*).

PRIAPULIDÆ.

Priapulus caudatus (Lam.).
Leith (*Colds.*).

DISCOPHORA.

HIRUDINEA.

Pontobdella muricata (Linn.).
Common as a parasite on the skate, and dredged in 12 fathoms, below Inchkeith (*nob.*).

MALACOBDELLEA.

Malacobdella grossa (O. F. Müll.).
Off Elie (*nob.*).
M. valenciennæi (Blanchard).
In *Mya truncata*, Firth of Forth (*Johnst. Cat.*).

POLYCHÆTA.

APHRODITIDÆ.

Aphrodite aculeata (Linn.).
Kirkcaldy Bay, 9 fathoms; Aberlady

Bay, 5 fathoms; shore, North Berwick, Newhaven, Portobello, etc. (*nob.*).

POLYNOIDÆ.

Lepidonotus squamatus (Linn.).
> Firth of Forth (*Brit. Mus.*); Portobello, 0-1 fathom, under stones (*Möb.*).
> This is a very common species on the oyster banks, and in all parts of the Firth.

Polynoë cirrata (Pallas).
> Portobello Sands, under stones (*Möb.*).

Halosydna gelatinosa (Sars).
> Off May Island, 8 fathoms (*nob.*).

SIGALIONIDÆ.

Sigalion idunæ (Rathke).
> Bass Rock, 24 fathoms (*Möb.*).

NEPHTHYDIDÆ.

Nephthys cœca (Fab.).
> Bass Rock, 24 fathoms (*Möb.*); beach, Aberdour, etc. (*nob.*).

N. johnstoni (Johnst.).
> Firth of Forth (*Brit. Mus.*).

PHYLLODOCIDÆ.

Phyllodoce laminosa (Sav.).
> Black Rocks, Leith (*Brit. Mus.*).

Eulalia viridis (O. F. Müll.).
> Off Longniddry, 14 fathoms, etc. (*nob.*).

NEREIDÆ.

Nereis pelagica (Linn.).
> Abundant in all parts of the Firth.

Hediste diversicolor (O. F. Müll.).
> Leith shore (*Möb.*).

Alitta virens (Sars).
> Beach, near Aberdour (*nob.*).

GLYCERIDÆ.
Glycera alba (Müll.).
> Bass Rock, 24 fathoms (*Möb.*).

OPHELIIDÆ.
Ammotrypane aulogaster (H. Rathke).
> Bass Rock, 24 fathoms (*Möb.*).

Ophelia acuminata (Œrst.).
> Firth of Forth (*Brit. Mus.*).

TELETHUSIDÆ.
Arenicola marina (Linn.).
> Very abundant in sand at low water mark.

Ephesia gracilis (Rathke).
> Firth of Forth (*Brit. Mus.*).

Trophonia plumosa (M. Edwards).
> Firth of Forth (*Brit. Mus.*); Wardie, between tide marks, Aberlady Bay, 9 fathoms; North Channel, 18 fathoms, etc. (*nob.*).

T. glauca (Malmgren).
> Off St Abb's Head, 40 fathoms (*Möb.*).

SPIONIDÆ.
Cirratulus cirratus (O. F. Müll.).
> Portobello (*Möb.*).

HERMELLIDÆ.
Sabellaria anglica (Ellis).
> Granton,—tubes only (*Brit. Mus.*).

AMPHICTENIDÆ.
Pectinaria belgica (Pallas).
> Firth of Forth (*Brit. Mus.*).

TEREBELLIDÆ.
Lanice conchilega (Pallas).
> Common in the upper laminarian zone all along the shores. Dredged in 18, 9, and 5 fathoms (*nob.*).

Thelepus circinatus (Fab.).
> Firth of Forth, 30 fathoms; Bass Rock, 24 fathoms (*Möb.*).

SABELLIDÆ.

Sabella pavonina (Sav.).
> Very common on the oyster bank, and in deeper water; also in sand at Aberdour at very low water.

SERPULIDÆ.

Serpula vermicularis (Linn.).
> Common, off Inchkeith, 5 fathoms, etc. (*nob.*).

Spirorbis borealis (Davd.).
> Very plentiful on Algæ at low water.

S. lucidus (Mont.).
> On Hydroids, etc., in all parts of the estuary.

PYCNOGONIDA.

This group has been recently separated from the CRUSTACEA and ARACHNIDA, with which it was formerly associated, and arranged as an independent class of the ARTHROPODA.

NYMPHONIDÆ.

Nymphon gracile (Leach).
> Not uncommon among stones and weeds between tide marks, and we have dredged it off May Island, 14 fathoms; Kirkcaldy Bay, 6 fathoms; off Inchkeith, 5 fathoms; and in many other parts of the estuary.

PHOXICHILIDÆ.

Pycnogonum litorale (O. F. Müll.).
> Common between tide marks, but also found in deep water. We have dredged it in 18 fathoms. We found specimens near Inchcolm, at the end of April of this year, having large masses of ova attached to them.

MOLLUSCA.

This list comprises the true Mollusca only. Of the groups which were formerly united under the title Molluscoida, and which are now generally regarded as members of the extensive and somewhat heterogeneous type Vermes, the Polyzoa and Tunicata have already been considered; while of the third, the Brachiopoda, there are no living representatives in the Firth of Forth.

On account of the comparatively advanced state of our knowledge of Molluscan faunas, and the special share of attention which is usually given to the group by collectors, we have reason to believe that this is more nearly complete than any of our other lists.

The nomenclature and arrangement are those given in Gwyn Jeffreys' well-known work.* We have frequently, however, mentioned also synonyms used by Forbes and Hanley ("A History of the British Mollusca," 1853), and other authors, especially when these were the names under which the species had been recorded from the Forth.

Dr M'Bain's extensive catalogue † was the foundation of the present list. The Mollusca of the German Expedition of 1872, which we have already had occasion to refer to several times, were reported on by Metzger and Meyer,‡ and we are indebted to their work for the localities of several species, five of which had not previously been recorded from the Firth of Forth.

Mr F. M. Balfour, F.R.S., has kindly given us a list of a few Mollusca which he dredged in the neighbourhood of Dunbar, including one of the rarest species in our list, *Pleurophyllidia lovéni*.

We desire to express our thanks to the Rev. J. M'Murtrie, M.A., for much valuable assistance. He kindly revised our MS., and from his notes added species, localities, and remarks, which, coming from one with his wide knowledge of British shells, and experience of the conchology of the Firth of

---

* "British Conchology," by J. Gwyn Jeffreys, F.R.S. London, 1863.
† "The East Neuk of Fife," by the Rev. Walter Wood. Edinburgh, 1862.
‡ Jahresb.—VIII., Mollusca.

Forth, were of very great importance to us. These contributions are subscribed (*M.*) in the list. As of peculiar interest we may mention his discovery of *Lepton squamosum* in shell sand at North Berwick. This species belongs to the south and south-west coasts of England, and the only Scotch locality mentioned in Jeffreys' "British Conchology" is Oban (Barlee). Mr M'Murtrie informs us that he has found the shell at Bamborough, and this, we believe, is the only place recorded where it has been found on the east coast.

LAMELLIBRANCHIATA.
ANOMIIDÆ.

Anomia ephippium (Linn.).
    Firth of Forth (*M'B.*); the so-called variety *aculeata*, which is common, and also var. *squamula* (young), at Granton, etc.; var. *cylindrica* on tangle stems at Leith and Granton (*M.*).
    This species is very common in the Firth. It occurs adhering to dead shells, etc., on the oyster bank; and is also found at low water mark attached to the roots and stems of *Laminaria*, etc.
    We have taken the varieties *cylindrica*, *aculeata*, and *squamula* at Aberdour.

A. patelliformis (Linn.).
    On North Berwick shore (*M.*).
    We have taken this species in from 5 to 7 fathoms, off Inchkeith.

A. patelliformis, var. *striata*.
    On Dunbar shore (*M.*); Firth of Forth (*M'B.*).
    Both *A. patelliformis* and the variety *striata* are common on roots of tangle cast up by storms at Newhaven (*M.*).

OSTREIDÆ.

Ostrea edulis (Linn.).
    Firth of Forth (*M'B.*, etc.).

The oyster occurs chiefly in the Firth in a few fathoms of water (5 to 10) on the bank stretching west from Inchkeith, and is dredged for the market by the fishermen from Newhaven, etc. We regret to say it is by no means plentiful.

The forms (they are scarcely varieties) *hippopus* and *deformis* occur (*M.*).

PECTINIDÆ.

Pecten pusio (Linn.).

Firth of Forth (*M'B.*).
Taken living in a tangle root, after a storm, on Craigroyston shore—attached by a byssus. Single valves are not uncommon on North Berwick shore, etc. (*M.*).

P. varius (Linn.).

Firth of Forth (*M'B.*).

P. opercularis (Linn.).

This is probably the commonest Lamellibranch in the Firth. It is abundant on the so-called "oyster bank" and other localities, and is dredged in large quantities by the Newhaven fishermen.

Specimens, with pure white shells, are brought in alive by storms somewhat plentifully at Craigroyston (*M.*).

Gwyn Jeffreys states that specimens of this species from the Firth of Forth "are much larger than usual," and that a specimen in his collection from Portobello measures $3\frac{2}{10}$ths inches long, and nearly 4 broad.

The variety *lineata*, with intermediate forms, is often taken by the Newhaven fishermen (*M.*).

P. septemradiatus? (Müll.).

It is probably this species which is mentioned in Fleming's "British Animals" under the name of *Pecten glaber* as being rare in the Firth of Forth.

Pecten tigrinus (Müll.).

Firth of Forth (*M'B.*, and *Ed. Mus.*).

Taken alive in Newhaven Harbour. It had probably been thrown away by the fishermen (*M.*).

We have dredged this species in the Firth, and have found it cast ashore at Elie.

P. tigrinus, var. *costata*.

Single valves of this, and also of the species, are found on the North Berwick shore (*M.*).

P. similis (Laskey).

Fifeshire (*Flem.*); Firth of Forth (*Laskey*).

P. maximus (Linn.).

Firth of Forth (*M'B.*).

Single valves are cast up at Craigroyston, but not plentifully (*M.*).

We dredged this species last summer off the Isle of May, in 8 fathoms.

MYTILIDÆ.

Mytilus edulis (Linn.).

Firth of Forth (*M'B.*, etc.).

Extremely common between tide marks and at low water on many parts of the shore. It occurs in profusion on the muddy beach at Wardie and Newhaven, where it is collected by the fishermen.

M. edulis, var. *incurvata*.

Abundant on sloping walls between Granton and Leith, wedged in between the stones (*M.*).

M. edulis, var. *pellucida*.

Granton (*M.*).

We have taken this variety at Wardie.

M. edulis, var. *galloprovincialis*.

Cast up on Craigroyston shore (*M.*).

Mytilus modiolus (Linn.).

Firth of Forth (*M'B.*, and *Ed. Mus.*).
Common at North Berwick, living at very low water, and in rock pools between tides. It is also frequently cast up alive along the whole coast. Large specimens are found in Newhaven Harbour, brought in by the fishermen (*M.*).

This species is frequently met with in dredging in the Firth. We have taken it to the east of Inchkeith, in 18 fathoms; west of Inchkeith, in 5 fathoms; off Fidra, in 7 fathoms; and north-west of the Isle of May, in 26 fathoms.

M. adriaticus (Lamk.).

Firth of Forth (*E. F.*).

Modiolaria marmorata (Forb.).

Off Elie (*M'B.*); Firth of Forth (*Ed. Mus.*). Plentiful on Newhaven shore in roots of *Laminaria*, after storms (*M.*).

Imbedded in the test of *Ascidia virginea*, not uncommon in 5 to 10 fathoms, especially on the bank west of Inchkeith. We have also taken it at low water at Wardie.

M. discors (Linn.).

Firth of Forth (*M'B.*, and *Ed. Mus.*).
Living with *M. marmorata* in roots of *Laminaria* on Newhaven shore after storms.

We have dredged this species in the Firth, and have also collected it at Portobello.

M. nigra (Gray).

Off Dudgeon (*Thomas*); Firth of Forth (*Ed. Mus.*, and *M'B.*).
I have a specimen from Mr Damon, Weymouth, marked "Black Rocks, Leith." I have sought for it at the lowest spring tides, without success (*M.*).

Crenella decussata (Mont.).
>Firth of Forth (*J. G. J.*).
>Single valves, and occasionally a perfect specimen, in drifted shell sand at North Berwick (*M.*).

ARCIDÆ.

Nucula nucleus (Linn.).
>Firth of Forth (*M'B., and Ed. Mus.*).
>We have dredged this species off Largo Bay, and near Inchkeith, in 7 fathoms.

N. nitida (Sowerby).
>Single valves of *Nucula* on this shore are usually of this species. I have taken perfect specimens (which, however, did not contain the animal) at very low water, Craigroyston, and at Cramond Island (*M.*). We have dredged this species off Elie, and west of Inchkeith, in 5 fathoms, and have also collected it on the beach at Largo Bay.

N. tenuis (Mont.).
>Firth of Forth (*J. G. J.*).

Leda minuta (Müll.).
>Firth of Forth (*M'B.*, under the name of *Leda caudata*).
>Small single valves occur in the drifted shell sand at North Berwick. The posterior extension is short. They are either the var. *brevirostris*, or the young state of the species (*M.*).

Pectunculus glycimeris (Linn.).
>Firth of Forth (*M'B.*).
>Occasionally in Newhaven Harbour, doubtless brought in by the fishermen (*M.*).

Arca tetragona (Poli).
>Firth of Forth, rare (*Flem.*, under the name of *Arca fusca*).

KELLIIDÆ.

Montacuta substriata (Mont.).
> On *Spatangus purpureus*, off Inchkeith (*Com. Mar. Zool.*); off the Isle of May, on *Spatangus purpureus*, 1854 (*M'B.*).

M. bidentata (Mont.).
> Single valves on Edinburgh shores. Perfect specimens not uncommon in drifted shell sand at North Berwick (*M.*).

M. ferruginosa (Mont.).
> Firth of Forth (*F. and H., and M'B.*).
> Single valves in drifted shell sand at North Berwick (*M.*).

Lasaea rubra (Mont.).
> North Berwick shore; not a plentiful shell here (*M.*).

Kellia suborbicularis (Mont.).
> In drifted shell sand at Dunbar and North Berwick. I have also taken it living at North Berwick in tangle root (*M.*).
> We have dredged this species in the Firth.
> Mr M'Murtrie informs us that at Alnmouth, Northumberland, he has taken both the species and the variety *lactea*, living in soft stone in rock pools at low water.

LUCINIDÆ.

Lucina borealis (Linn.).
> Firth of Forth (*M'B.*).
> Occasionally on the beach at Newhaven, North Berwick, etc. (*M.*); Elie (*nob.*).

Axinus flexuosus (Mont.).
> Firth of Forth (*M'B.* as *Lucina flexuosa*), Cramond Island, and adjoining shore,

chiefly single valves, but sometimes a perfect specimen (*M.*).

We have collected this species at Aberdour, etc.

CARDITIDÆ.

Cyamium minutum (Fabr.).

Living among small seaweeds in rock pools, etc., at North Berwick and Dunbar. Very common at the Isle of May. A clear white form is rare at North Berwick (*M.*).

CARDIIDÆ.

Cardium echinatum (Linn.).

Firth of Forth (*M'B., and Ed. Mus.*); Bass Rock, 24 fathoms, and St Abb's Head 40 fathoms (*Metz. and Mey.*).

Taken alive between Granton and Cramond Island. Specimens with the prickles worn smooth are not uncommon, and bear a great resemblance to *C. tuberculatum*, and have been mistaken for it. Good prickly specimens are frequent in Newhaven Harbour, brought in by the fishermen. Single valves occur along the whole coast (*M.*).

This species is not uncommon in the Firth. We have dredged it off Longniddry, in 7 fathoms; west of Inchkeith, in 12 fathoms; and have collected it on the beach at Largo and Portobello.

C. fasciatum (Mont.).

We have a specimen of this species from Elie.

C. edule (Linn.).

Firth of Forth (*M'B., and Ed. Mus.*).

This species is common. We have taken it at Cramond, Newhaven, Elie, North Berwick, Aberdour, etc.

Cardium norvegicum (Spengler).
>Firth of Forth (*M'B.*); off Inchkeith (*Com. Mar. Zool.*).
>We have dredged this species off the Isle of May, in 8 fathoms, and near Portobello.

CYPRINIDÆ.

Cyprina islandica (Linn.).
>Firth of Forth (*M'B., and Ed. Mus.*); Firth of Forth, very low tides (*F. and H.*).
>I have taken it alive, and apparently *in situ*, in muddy sand, between Granton and Newhaven at very low water (*M.*).
>We have dredged this species east of Inchkeith, 18 fathoms; north-west of the Isle of May, 26 fathoms; and have found it frequently on the beach at Portobello, etc.

Astarte sulcata (Da Costa).
>Firth of Forth (*M'B.*).
>We have dredged this species in the Firth.

A. sulcata, var. *elliptica*.
>Firth of Forth, dead valves (*M'B.*).

A. compressa (Mont.).
>Firth of Forth, not rare on the oyster banks in 7 to 14 fathoms (*F. and H.*); Firth of Forth (*M'B.*).

A. compressa, var. *striata*.
>Firth of Forth, plentiful (*J. G. J.*).
>A living specimen cast up on the beach in Canty Bay. Single valves are common on the shore between Canty Bay and North Berwick (*M.*).
>We have dredged this species near Inchkeith, in 7 fathoms.

A. triangularis (Mont.).
>Single valves, and sometimes a perfect specimen, in drifted shell sand at North Berwick (*M.*).

VENERIDÆ.

Venus exoleta (Linn.).
Firth of Forth (*M'B.*); Firth of Forth, 7 fathoms (*F. and H.*).
Frequently cast up alive at North Berwick; common at Newhaven, brought in by fishermen (*M.*).
We have dredged it west of Inchkeith in 5 fathoms, and at Newhaven.

V. lincta (Pult.).
Firth of Forth (*M'B., and F. and H.*); Bass Rock, 24 fathoms (*Metz. and Mey.*).
Cast up alive at Leven, Fife, and on the North Berwick shore, near Fidra. It appears to be less common in this Firth than it is in St Andrew's Bay (*M.*).
We have dredged it in 7 fathoms.

V. fasciata (Da Costa).
Firth of Forth (*M'B.*).
Cast up on the shore at North Berwick (*M.*).
We have taken this species on the shore at Anstruther.

V. fasciata, var *radiata*.
Cast up on the shore at North Berwick (*M.*).

V. casina (Linn.).
Dead valves occasionally at North Berwick, especially on the east shore (*M.*).

V. ovata (Penn.).
Firth of Forth (*M'B.*).
Dead valves at Cramond Island and Canty Bay (*M.*).

V. gallina (Linn.).
Firth of Forth (*M'B., and Ed. Mus.*); Bass Rock, 24 fathoms (*Metz. and Mey.*).
Alive on Cockenzie shore at low water, sometimes without any markings on the shell (*M.*).

We dredged this species last summer in Aberlady Bay in 9 fathoms, and have found it alive on the beach at Portobello, Aberdour, etc.

Venus gallina, var. *gibba*.

Between Granton and Cramond Island, with intermediate forms (*M.*).

Tapes virgineus (Linn.).

Firth of Forth (*Ed. Mus.*, and *M'B.*).

Good specimens in Newhaven Harbour, sometimes milk-white (*M.*).

We have found this species at Portobello.

T. pullastra (Mont.).

Firth of Forth; plentiful at low water, especially near Newhaven and Cramond (*F. and H.*); Firth of Forth (*Ed. Mus., and M'B.*); Dalmeny, where it grows to a large size—$2\frac{1}{4}$ inches broad—occasionally pure white (*M.*); beach between Portobello and Fisherrow (*Metz. and Mey.*).

We dredged it last summer off the Isle of May, in 8 fathoms, and have collected it at Elie, Cramond, and Portobello.

A monstrosity, showing the foliaceous structure of the shell, gathered at Cockenzie, has the front margin, as it were, of three shells, but only one hinge-line (*M.*).

T. pullastra, var. *perforans*.

Living in soft rock at Wardie and North Berwick at low water. Semi-fossil at Longniddry and other places (*M.*).

T. decussatus (Linn.).

Beach between Portobello and Fisherrow (*Metz. and Mey.*).

Lucinopsis undata (Penn.).

Firth of Forth (*M'B., F. and H., and Ed. Mus.*); Bass Rock, 24 fathoms (*Metz. and Mey.*).

Cast up by storms at Portobello, Granton, etc. (*M.*).

We have dredged this species in Largo Bay.

TELLINIDÆ.

Tellina crassa (Gmelin).
Firth of Forth (*M'B.*, *and Knapp*).
Single valves common at North Berwick, especially on the east shore near the harbour (*M.*).

T. balthica (Linn.).
Firth of Forth (*M'B.*, as *T. solidula*); Firth of Forth (*Ed. Mus.*); Common at Leith, etc., living between tide-marks (*M.*).
We have collected this species in Largo Bay, Aberdour, etc.

T. tenuis (Da Costa).
Firth of Forth (*M'B., and Ed. Mus.*).
Common at Granton, Portobello, etc., living at low water, and brought up by storms (*M.*).
We have found this species in Largo Bay.

T. fabula (Gronovius).
Firth of Forth (*M'B.*); with the preceding, common at Granton, Cramond Island, Portobello, etc. (*M.*).
We have collected it at Portobello, and in Largo Bay.

Psammobia tellinella (Lamk.).
Single valves on the beach at North Berwick and Canty Bay, but not common (*M.*).
We have collected this species in Largo Bay.

P. ferröensis (Chemn.).
Bass Rock, 24 fathoms (*Metz. and Mey.*); Firth of Forth (*M'B., F. and H., and Ed. Mus.*); on Craigroyston shore and Dirleton shore, brought up by storms (*M.*).

We have dredged it in Aberlady Bay, 9 fathoms, and off Leith.

Psammobia vespertina (Chemn.).
Firth of Forth (M'B., and J. G. J.).

Donax vittatus (Da Costa).
Firth of Forth (M'B.); Craigroyston, Aberlady, and Dirleton shores (M.).
We have found it at Elie and Largo.

MACTRIDÆ.

Mactra solida (Linn.).
Beach at Fisherrow (Metz. and Mey.); Firth of Forth (Ed. Mus., and M'B.).
We have dredged it off Elie.

M. solida, var. *truncata*.
Firth of Forth (M'B., and J. G. J.); Firth of Forth, 7 fathoms (F. and H.). Lives with the species, and with intermediate forms, on the shore from Newhaven to Cramond Island (M.).
We have obtained this variety on Cramond Island.

M. solida, var. *elliptica*.
Firth of Forth (M'B., and F. and H.).

M. subtruncata (Da Costa).
Firth of Forth (F. and H., and M'B.); Granton and Portobello (M.).
We have dredged it in the Firth.

M. subtruncata, var. *striata*.
Granton, not common (M.).

M. subtruncata, var. *inaequalis*.
Dead shells at South Queensferry (M.).

M. stultorum (Linn.).
Firth of Forth (M'B., and Ed. Mus.); Portobello (M.).
This species is sometimes found cast up in large quantities on Portobello sands after storms.

Mactra stultorum, var. *cinerea*.
> Mouth of the Almond, etc. (*M.*).

Lutraria elliptica (Lamk.).
> Firth of Forth (*M'B.*, *Ed. Mus.*, and *F. and H.*); Portobello (*M.*).
> We have collected this species at Portobello, and at Caroline Park.

Scrobicularia prismatica (Mont.).
> Firth of Forth (*F. and H.*); off Elie (*M'B.*). Occurs sparingly alive on the sands at very low water between Granton and Cramond Island, especially towards Cramond Island. Empty shells, with the valves still united, are a little commoner (*M.*).
> We have found this species on the shore west from Aberdour.

S. nitida (Müller).
> St Abb's Head, 40 fathoms (*Metz. and Mey.*).

S. alba (Wood).
> Firth of Forth (*Ed. Mus.*); Off Elie (*M'B.*). Frequent at Portobello, etc., living at low water, and on the beach brought up by storms (*M.*).

S. piperata (Bellon.).
> Firth of Forth (*F. and H., and M'B.*); Caroline Park, shells well preserved, but not living; they are in a bed of blue clay between tides (*M.*).
> We dredged this species last summer in 5 fathoms, to the south-west of Inchkeith.

SOLENIDÆ.

Solen pellucidus (Penn.).
> Firth of Forth (*M'B.*, and *Thomas and Knapp*); brought up by the tides between Portobello and Leith, and between Granton and Cramond Island, but not plentiful (*M.*); east of Inchkeith (*J. Hunter Bar-*

ron); Bass Rock, 24 fathoms (*Metz. and Mey.*, as *Cultellus pellucidus*).

Solen ensis (Linn.).
Firth of Forth (*M'B., and Ed. Mus.*).
We have taken *Solen ensis* frequently.

S. siliqua (Linn.).
Firth of Forth (*M'B., and Ed. Mus.*).
We dredged it lately west of Inchkeith, in 4 fathoms.

S. siliqua, var. *arcuata*.
Newhaven shore (*M.*).

ANATINIDÆ.

Thracia prætenuis (Pult.).
Firth of Forth (*M'B., and F. and H.*).

T. papyracea (Poli).
Firth of Forth (*Ed. Mus., and M'B.*).
We have taken this species frequently at Portobello, Cramond, etc.

T. convexa (W. Wood).
Firth of Forth, dead shells (*M'B.*).
Single valves not unfrequent on the shore near Edinburgh, at low water, especially towards Cramond Island (*M.*).

T. distorta (Mont.).
Firth of Forth (*F. and H., and M'B.*).

CORBULIDÆ.

Neæra cuspidata (Olivi).
Off Portseaton and Fidra, 17 fathoms (*Thomas*); Firth of Forth (*F. and H., and M'B.*).

Corbula gibba (Olivi).
Firth of Forth (*Ed. Mus., M'B., and F. and H.*).
Single valves are frequent on the shores at North Berwick, Newhaven, Granton, and South Queensferry. I have taken a

perfect specimen (young) on the shore at Cramond Island (*M.*).

We have dredged this species frequently in the Firth — in Aberlady Bay, 9 fathoms; west of Inchkeith, 5 to 7 fathoms, etc.

MYIDÆ.

Mya arenaria (Linn.).

Firth of Forth (*M'B., Ed. Mus., and F. and H.*); Bass Rock, 24 fathoms (*Metz. and Mey.*); Granton, living at low water west from Leith pier (*M.*).

We have found this species in Largo Bay.

M. arenaria, var. *lata*.

Biel sands, Longniddry, and Newhaven. Rather common in west harbour, Granton, (*M.*).

We have found this variety at Aberdour.

M. truncata (Linn.).

Firth of Forth, from low water to 7 fathoms (*F. and H.*); Firth of Forth (*M'B.*). Living at Granton, Craigroyston, and Cramond Island (*M.*).

We have found this species at Portobello, and have dredged it in 4 fathoms, to the west of Inchkeith.

M. binghami (Turt.).

Firth of Forth (*F. and H., and M'B.*).

SAXICAVIDÆ.

Saxicava rugosa (Linn.).

Firth of Forth (*M'B.*); Firth of Forth, 7 fathoms (*F. and H.*).

Taken living, and apparently at home, in pure sand, between Granton and Cramond Island (*M.*).

This species is very common from the upper laminarian zone downwards. We have dredged it east of Inchkeith, in 18 fathoms; off Longniddry, in 14 fathoms; west of Inchkeith, in 5 fathoms; and have collected it at Aberdour, Elie, and Largo.

Saxicava rugosa, var. *minuta*.

Granton, North Berwick, and Dunbar (*M.*). We have some specimens of this variety from Largo Bay.

S. rugosa, var. *arctica*.

Firth of Forth (*M'B., and Cunningham*). On tangle roots at Granton, etc. (*M.*).

GASTROCHÆNIDÆ.

Gastrochæna dubia (Penn.).

On Craigroyston beach, after a storm, I picked up a fossil coral, containing in cavities fresh but empty shells of *G. dubia* (*M.*).

PHOLADIDÆ.

Pholas dactylus (Linn.).

Firth of Forth (*M'B.*).

P. candida (Linn.).

Firth of Forth (*F. and H., Ed. Mus., and M'B.*); Newhaven, containing the animal (*M.*).

We have found dead valves at Cramond Island, Aberdour, and in Largo Bay.

P. crispata (Linn.).

Firth of Forth (*F. and H., M'B., and Ed. Mus.*); Granton, west of harbour, alive at very low water, also North Berwick (*M.*); near Cramond (*J. Hunter Barron*).

We have found this species alive at Wardie, and at Cramond, and have found dead valves at Largo and other places.

SOLENOCONCHIA.

DENTALIDÆ.

Dentalium entalis (Linn.).

Bass Rock, 24 fathoms (*Metz. and Mey.*);
Firth of Forth (*Ed. Mus., M'B.*, etc.).
Rare on the shore at Granton, Leith, and
North Berwick (*M.*).
We have dredged this species off Inchkeith, in 9 fathoms; off Largo Bay, etc.
It is often obtained by the fishermen
adhering to the long lines; dead specimens
are not uncommon on some parts of the
shore, as Largo Bay, Elie, and Aberdour.

GASTROPODA.

I. CYCLOBRANCHIATA—

CHITONIDÆ.

Chiton fascicularis (Linn.).

Firth of Forth (*M'B., F. M. B.*).
On tangle roots, Craigroyston shore, after
a storm. At North Berwick alive on
under side of stones in rock pools, at low
water, but not plentiful (*M.*).
We have taken this species at low water
mark, near Elie.

C. cinereus (Linn.).

Firth of Forth (*F. M. B., M'B.*, under the
name of *Chiton asellus*).
We have dredged this species frequently
in a few fathoms.

C. marginatus (Penn.).

Firth of Forth (*M'B.*); between tide marks
(*F. M. B.*); beach between Portobello and
Fisherrow (*Metz. and Mey.*).
Very large at end of East Pier, Leith
(near Martello Tower), at very low water.
Length of dried specimens 0·9 inch;
breadth, 0·55 (*M.*).

This species is common between tide marks at Wardie, Elie, and other parts of the Firth. Fleming, in his " British Animals," mentions having found a specimen at Newhaven, with only six valves.

Chiton ruber (Lowe).

Firth of Forth (*M'B.*); between tide marks (*F. M. B.*); abundant on roots and stems of *Laminaria* on North Berwick shore, after storms (*M.*).

C. marmoreus (Fabr.).

Rare on Black Rocks, Leith (*Knapp*).

We have taken this species in 5 fathoms, about half a mile to the west of Inchkeith.

II. PECTINIBRANCHIATA—

PATELLIDÆ.

Patella vulgata (Linn.).

Extremely common between tide marks on rocky shores. Large, heavy specimens at Cramond Island.

P. vulgata, var. *picta*.

Granton and South Queensferry (*M.*).

We have taken this variety on the rocks at Joppa. Specially common at North Berwick, at low water.

P. vulgata, var. *cærulea*.

Granton and North Berwick, not rare. It shades off, through intermediate forms, into var. *picta* (*M.*).

We have obtained it at Wardie.

P. vulgata, var. *depressa*.

The *Patella athletica* of Forbes and Hanley. This well-marked variety is frequent along the whole North Berwick coast, at very low water, and in rock pools between tides. It grows in its longest diameter to 2·3 inches. It becomes rare as we go up the Firth, being evidently not an estuary shell. I have not found it beyond Gullane shore, where I took a solitary living specimen. It is abundant at the Isle of May (*M.*).

Note.—Where a spring of fresh water rises on the shore above low water mark, a little to the east of Granton East Harbour, a thin form of *P. vulgata*, pale, and with a silky surface, takes the place of all other forms, and is plentiful. It may be due merely to the influence of fresh water; but I suspect a corroding quality in the spring. *Tectura testudinalis* in the same place is worn very thin. If this form of *P. vulgata* had permanence, its appropriate name would be var. *sericea* (*M.*).

Helcion pellucidum (Linn.).

Firth of Forth (*M'B.*).

Good specimens common on *Laminaria* all along the North Berwick coast, and at Dunbar and Cramond Island (*M.*).

We have found this species at extreme low water, at Elie, Aberdour, and North Berwick. It frequents the stems and roots of *Laminaria*.

H. pellucidum, var. *lævis*.

Common in the hollow part (or under side) of roots of *Laminaria* at North Berwick; also at Dunbar and Longniddry (*M.*).

We obtained the variety also at Elie and Aberdour, and on the beach in Largo Bay.

Tectura testudinalis (Müller).

Firth of Forth (*M'B.*); beach between Portobello and Fisherrow (*Metz. and Mey.*); Granton (*M.*).

The shell is sometimes an inch long on the Edinburgh shores. The markings vary greatly, and there is a blue form bearing the same relation to the species as the variety *cærulea* does to *Patella vulgata* (*M.*). This species is common on the shores of the Firth. We have taken it at Aberdour, Wardie, etc.

Tectura virginea (Müller).

Firth of Forth (*M'B., Ed. Mus.*); Dunbar, at North Berwick living on the rocks at very low water (*M.*).

FISSURELLIDÆ.

Emarginula fissura (Linn.).

Firth of Forth (*Ed. Mus.*); empty shells on North Berwick beach (*M.*).

We have found this shell on the beach at Largo Bay.

CAPULIDÆ.

Capulus hungaricus (Linn.).

Alive off Inchkeith, May 1861 (*M'B.*); off Inchkeith (*Com. Mar. Zool.*); small empty shells on North Berwick beach (*M.*).

We have dredged it alive several times in the Firth,—Aberlady Bay, 9 fathoms; and east of Inchkeith, 7 fathoms. We have also found it cast ashore on Portobello beach.

TROCHIDÆ.

Trochus helicinus (Fabr.).

Not rare in shell sand at North Berwick and Dunbar (*M.*).

T. magus (Linn.).

I have found a worn shell on Craigroyston beach, but I do not suppose that this species inhabits the Firth of Forth (*M.*).

T. tumidus (Mont.).

Firth of Forth (*M'B., and Ed. Mus.*).

We have taken this species in 7 fathoms, off Inchkeith.

Trochus cinerarius (Linn.).
> Firth of Forth (*M'B., and Ed. Mus.*); beach between Portobello and Fisherrow (*Metz. and Mey.*).
> This is a very common species in the Firth about low water mark. It is abundant at Granton, Wardie, Aberdour, Elie, and other parts of the coast. We have also dredged it off Inchkeith in 5 fathoms.

T. zizyphinus (Linn.).
> Firth of Forth (*M'B.*).
> Not common here. I have found dead shells on the beach at Edinburgh and Dirleton (*M.*).

LITTORINIDÆ.

Lacuna crassior (Mont.).
> Firth of Forth (*Ed. Mus., and M'B.*).
> Not rare on the beach at Wardie, Granton, and North Berwick (*M.*).

L. divaricata (Fabr.).
> Firth of Forth (*Ed. Mus., and M'B.*, as *L. vincta*).
> Common on *Laminaria*, etc., at Granton, North Berwick, Dunbar, and other places (*M.*).

L. divaricata, var. *canalis*.
> With the species at North Berwick, etc., somewhat common at Granton (*M.*).

L. divaricata, var. *quadrifasciata*.
> On the beach at Leven, Fife (*M.*).

L. pallidula (Da Costa).
> Firth of Forth (*Ed. Mus., J. G. J., and M'B.*).
> Common on *Laminaria*, etc., at Granton, North Berwick and Dunbar (*M.*).
> We found this species in Largo Bay.

Littorina obtusata (Linn.).
> Firth of Forth (*Ed. Mus., and M'B.*);

Granton (*M.*); beach between Portobello and Fisherrow (*Metz. and Mey.*).

Very common in the littoral zone at Wardie, Aberdour, Elie, etc.

Littorina obtusata, var. *fabalis*.

Not uncommon at Dunbar, etc.; a doubtful variety (*M.*).

L. neritoides (Linn.).

Firth of Forth (*M'B.*).

North Berwick, on the harbour rocks, above high water mark. Abundant and well grown at the landing-place on the Isle of May (*M.*).

L. rudis (Maton).

Firth of Forth (*Ed. Mus.*, and *M'B.*); Granton (*M.*); beach between Portobello and Fisherrow (*Metz. and Mey.*); very common at Wardie, Elie, Aberdour, etc.

L. rudis, var. *saxatilis*.

North Berwick, living with *L. neritoides*; and dead in shell sand (*M.*).

L. rudis, var. *lincata*.

Granton. Is this a variety, or a common form? (*M.*).

Common at Wardie.

L. rudis, var. *tenebrosa*.

Granton and Dunbar (*M.*).

L. littorea (Linn.).

Firth of Forth (*M'B.*, and *Ed. Mus.*); Granton, sometimes orange-red (*M.*).

Very common on the beach at Wardie, etc.

Rissoa reticulata (Mont.).

We have found this species on the shore, near Aberdour.

R. punctura (Mont.).

Also found near Aberdour.

Common in the shell sand at North Berwick and Dunbar (*M.*).

Rissoa costata (Adams).

Common in the shell sand at North Berwick and Dunbar (*M.*).

R. parva (Da Costa).

Firth of Forth (*M'B.*).
Not so common as the variety *interrupta*. Good specimens at North Berwick and Dunbar (*M.*).

R. parva, var. *interrupta*.

Very common on seaweeds at low water at Granton, North Berwick, etc. (*M.*).
We have found this variety on the shore near Aberdour.

R. striata (Adams).

Firth of Forth (*M'B.*).
Alive among small seaweeds in rock pools at low water, North Berwick. Common in shell sand at Dunbar, North Berwick, Belhaven, Granton, etc. (*M.*).
We have found it in sand at Elie.

R. striata, var. *arctica*.

Occurs with the species, but is less common (*M.*).

R. vitrea (Mont.).

Rare in the shell sand at North Berwick (*M.*).
It has been found at Dunbar by Bingham.

R. semistriata (Mont.).

Firth of Forth, 7 fathoms (*F. and H.*).
Common in the shell sand at North Berwick (*M.*).

R. cingillus (Mont.).

North Berwick, in the shell sand, but not plentiful (*M.*).
We obtained this species on the shore near Largo.

Hydrobia ulvæ (Penn.).

Alive at Pefferburnfoot, Aberlady; dead on the beach at Cramond, Newhaven, North

Berwick, etc. Living abundantly on mud flats at Bo'ness (*M.*).

White specimens of *H. ulvæ* come ashore at Newhaven. As they are dead, and may be only bleached, it would be unsafe to say that they are the var. *albida* (*M.*).

We have found this species in Largo Bay.

SKENEIDÆ.

Skenea planorbis (Fabr.).

Firth of Forth (*M'B.*).

Abundant on *Cladophora* and other seaweeds at Dunbar, North Berwick, and Granton (*M.*); Elie (*nob.*).

Homalogyra atomus (Philippi).

In shell sand between North Berwick and Canty Bay, not plentiful (*M.*).

VERMETIDÆ.

Cæcum trachea (Mont.).

Firth of Forth (*Ed. Mus.*).

C. glabrum (Mont.).

Not rare in fine shell sand at North Berwick (*M.*).

TURRITELLIDÆ.

Turritella terebra (Linn.).

Firth of Forth (*M'B.*, as *T. communis*); Firth of Forth, 30 fathoms (*Metz. and Mey.*, as *T. ungulina*); Firth of Forth (*Ed. Mus.*). Alive on the shore at very low water, near Cramond Island (*M.*).

This species is very common in the Firth; in some spots—*e.g.*, off the east side of Inchkeith, 12 fathoms; off Aberdour, 5 fathoms; and Kirkcaldy Bay, 9 fathoms—the dredge comes up almost full of dead shells of this species, imbedded in slimy grey mud.

Turritella terebra, var. *nivea*.
> Not rare on the shore, both east and west from Edinburgh. Specially common towards Cramond Island, and on the shore east from Burntisland (*M.*).

T. terebra, var. *gracilis*.
> Granton and towards Cramond Island, not common (*M.*).

PYRAMIDELLIDÆ.

Aclis unica (Mont.).
> Somewhat rare in fine shell sand on North Berwick beach (*M.*).

Odostomia rissoïdes (Hanley).
> In shell sand, North Berwick (*M.*).

O. rissoïdes, var. *alba*.
> Shell sand, North Berwick (*M.*).

O. rissoïdes, var. *nitida*.
> Shell sand, North Berwick (*M.*).

O. rissoïdes, var. *dubia*.
> Shell sand, North Berwick. This species is highly variable, and forms occur which it is difficult to assign to any of these varieties (*M.*).

O. pallida (Mont.).
> Not unfrequent in the shell sand at North Berwick (*M.*).

O. acuta (Jeffreys).
> Occasionally in the shell sand, North Berwick (*M.*).

O. acuta, var. *umbilicata*.
> Not unfrequent in the shell sand, North Berwick (*M.*).

O. unidentata (Mont.).
> Somewhat common in the shell sand at North Berwick and Dunbar (*M.*).

O. decussata (Mont.).
> Somewhat rare in shell sand, North Berwick (*M.*).

Odostomia indistincta (Mont.).
: Not rare in shell sand, North Berwick (*M.*).

O. indistincta, var. *brevior.*
: In shell sand, North Berwick (*M.*).

O. interstincta (Mont.).
: Common in shell sand at North Berwick and Dunbar (*M.*).

O. spiralis (Mont.).
: Common in shell sand at North Berwick and Dunbar (*M.*).

O. acicula (Philippi), var. *ventricosa.*
: Rare in the shell sand at North Berwick. My specimens do not depart widely from the species, *i.e.*, the whorls are but slightly ventricose (*M.*).

EULIMIDÆ.

Eulima distorta (Deshayes).
: Rare in fine shell sand, North Berwick (*M.*).

E. bilineata (Alder).
: Not rare in shell sand at North Berwick and Dunbar (*M.*).

NATICIDÆ.

Natica catena (Da Costa).
: Firth of Forth (*Ed. Mus.,* and *M'B.*).
: Alive on Leith Sands, at very low water; cast up at Joppa, alive; Longniddry (*M.*).

N. alderi (Forb.).
: Firth of Forth (*M'B.*); Bass Rock, 24 fathoms (*Metz.* and *Mcy.*).
: Dead shells on the beach at Elie, Craigroyston (worn), and North Berwick (*M.*).
: We have taken this species in Largo Bay.

N. alderi, var. *lactea.*
: North Berwick (*M.*).
: Along with the species in Largo Bay.

N. islandica (Gmel.).
: Haddocks' stomachs, Firth of Forth (*Dr Knapp*).

VELUTINIDÆ.

Lamellaria perspicua (Linn.).
>North Berwick in shell sand, small and not abundant (*M.*).

Velutina lævigata (Penn.).
>Firth of Forth (*Ed. Mus.*); on the beach at Dunbar, North Berwick, Longniddry, Granton, and Cramond (*M.*).
>This species is not uncommon in a few fathoms of water. We have dredged it in Kirkcaldy Bay, 9 fathoms; off Inchkeith, 5 fathoms; and in Aberlady Bay, 9 fathoms. We have also collected it dead on the beach at Largo Bay.

APORRHAIDÆ.

Aporrhais pes-pelecani (Linn.).
>Firth of Forth (*M'B.*); off Inchkeith (*Com. Mar. Zool.*); Firth of Forth, 30 fathoms (*Metz. and Mey.*).
>Brought in alive by the Newhaven fishermen. On the beach at North Berwick and Craigroyston (*M.*).
>Not uncommon. We have dredged it in Largo Bay and off Kirkcaldy, and have also obtained it on the beach at Largo.

CERITHIIDÆ.

Cerithium reticulatum (Da Costa).
>A worn fragment on the beach at North Berwick. I do not think the species lives on this coast. However I found a perfect though dead shell on the Northumbrian shore at Alnmouth (*M.*).

III. SIPHONOBRANCHIATA—

BUCCINIDÆ.

Purpura lapillus (Linn.).
>Firth of Forth (*M'B.*).

This species is extremely common in the littoral zone in the Firth.

Purpura lapillus, var. *imbricata*.

Granton and Newhaven. I have not found a good specimen of this variety at North Berwick, though the species varies very much there as regards colour (*M.*); Firth of Forth (*Dr Knapp*).

This variety is common at Wardie, Aberdour, etc., between tide marks.

Buccinum undatum (Linn.).

Firth of Forth (*Ed. Mus., and M'B.*).

Forms occur which approach the varieties *littoralis* and *striata*, but they are not very well marked (*M.*).

Very common on the oyster bank and in other parts of the Firth. We have also taken it between tide marks at Wardie and Aberdour.

MURICIDÆ.

Murex erinaceus (Linn.).

Firth of Forth (*Ed. Mus.*).

Dead shells of *M. erinaceus* are not rare on the beach at North Berwick. I took a living specimen in a rock pool at very low water there (*M.*).

Trophon truncatus (Ström.).

Wardie, Portobello, and North Berwick (*M.*); Firth of Forth (*Ed. Mus.*).

We have dredged this species in the Firth, and have also taken it at Wardie.

Fusus antiquus (Linn.).

Firth of Forth (*Ed. Mus.*); Firth of Forth, 30 fathoms (*Metz. and Mey.*); Off the Isle of May (*M'B.*).

This species is not uncommon. We have dredged it in Kirkcaldy Bay, 9 fathoms; north-east of Inchkeith, 12 fathoms; and between Inchkeith and the Isle of May, 18 fathoms. Often brought to Newhaven pier in the fishing boats.

Monstr. *varicosum*, with several persistent outer lips, whitish, and large (6¼ inches long). Newhaven Harbour, brought in by the fishermen (*M.*).

Fusus gracilis (Da Costa).

Firth of Forth, 30 fathoms (*Metz. and Mey.*). We have obtained it from the fishing boats at Newhaven. Probably it is this species which is referred to by M'Bain under the name of *Fusus islandicus*.

F. propinquus (Alder).

Firth of Forth, 30 fathoms (*Metz. and Mey.*); off the Isle of May, 1854 (*M'B.*).

F. jeffreysianus (Fischer).

Firth of Forth, 30 fathoms (*Metz. and Mey.*).

NASSIDÆ.

Nassa reticulata (Linn.).

Firth of Forth (*M'B.*).

N. incrassata (Ström.).

Firth of Forth (*M'B., and Ed. Mus.*). At Wardie it is commoner than the var. *minor* (*M.*).

This species is very common in the Firth. We have taken it in abundance at North Berwick, Wardie, Aberdour, etc.

N. incrassata, var. *minor*.

Living, and abundant, on the under side of stones and crumbling rocks at North Berwick, at very low water, and in roots of tangle. This variety is commoner than the species at North Berwick and Elie. It is merely a dwarf form, with a fully developed labial rib. Children at North Berwick pierce them with needle and thread, and call them necklace shells (*M.*).

Nassa incrassata, var. *simulans*.
>Occasionally at North Berwick and Granton (*M.*).

PLEUROTOMIDÆ.

Defrancia linearis (Mont.).
>Common on the beach at North Berwick and Dunbar (*M.*).

D. linearis, var. *aequalis*.
>Somewhat common on the beach at North Berwick and Dunbar (*M.*).

Pleurotoma costata (Donov.).
>Firth of Forth; in shell sand at Granton, Longniddry, North Berwick, and Dunbar (*M.*).
>We have found this species on the beach near Aberdour.

P. septangularis (Mont.).
>Firth of Forth (*M'B.*).

P. rufa (Mont.).
>Wardie and North Berwick, in shell sand (*M.*).

P. turricula (Mont.).
>Firth of Forth (*M'B.*); on the beach at Wardie, Longniddry, North Berwick, and Dunbar (*M.*); Bass Rock, 24 fathoms (*Metz. and Mey.*).

P. turricula, var. *rosea*.
>North Berwick and Dunbar (*M.*).

P. trevelyana (Turt.).
>Bass Rock, 24 fathoms (*Metz. and Mey.*); Firth of Forth (*M'Andrew*).

CYPRÆIDÆ.

Marginella laevis (Donov.).
>Laskey gave Dunbar as a locality, but his specimen was a tropical species (*Dr Gwyn Jeffreys*).

I have a young shell (two-thirds grown) which I took alive from the under side of a stone in a rock pool, at very low water, opposite the Marine Hotel, North Berwick. It is thin and semi-transparent, with a sharp outer lip, the young of this species being unlike the mature form. I have gathered the shell at the same stage of growth on Herm beach, near Guernsey, where *M. laevis* is not rare (*M.*).

Cypræa europæa (Mont.).
Largo Bay, etc. (*M'B.*); Firth of Forth (*Ed. Mus.*).
Rare on the Edinburgh beach. It becomes more frequent as we go east, and at North Berwick is sometimes very abundant. At North Berwick adult specimens worn smooth, and having about two whorls of the spire laid bare, are not uncommon (*M.*). We have taken it at North Berwick, Largo Bay, Anstruther, etc.

IV. PLEUROBRANCHIATA—

BULLIDÆ.

Cylichna umbilicata (Mont.).
Firth of Forth (*M'B.*).
C. cylindracea (Penn.).
Bass Rock, 24 fathoms (*Metz. and Mey.*); Firth of Forth (*M'B.*); on the beach at Dirleton and North Berwick, but not plentiful (*M.*).
We have found this species in Largo Bay.
Utriculus truncatulus (Brug.).
Firth of Forth (*M'B.*); common on the beach at North Berwick and Dunbar (*M.*). We have found it on the beach west from Aberdour.

Utriculus truncatulus, var. *pellucida*.
> Occasionally at North Berwick and Dunbar (*M.*).

U. obtusus (Mont.).
> Firth of Forth (*M'B.*, and *J. G. J.*); on the beach at Craigroyston (*M.*).
> Along with the last species near Aberdour.

U. hyalinus (Turton).
> Not rare in shell sand at Cramond Island, North Berwick, and Dunbar (*M.*).

Actæon tornatilis (Linn.).
> Firth of Forth (under the name of *Tornatella fasciata*, M'B.); on the beach at Edinburgh, Portobello, and Dirleton, but not common (*M.*).
> We have taken this shell at Wardie and Aberdour.

Philine catena (Mont.).
> Bass Rock, 24 fathoms (*Metz. and Mey.*).
> Somewhat common in the shell sand at North Berwick and Dunbar (*M.*).

P. pruinosa (Clark).
> Firth of Forth (*Flem.*).

P. aperta (Linn.).
> Firth of Forth (*M'B., Ed. Mus., and Forb.*); alive at very low water at Portobello and near Cramond Island, sometimes plentiful (*M.*).
> We have taken this species alive on the Silver Sands, Aberdour; and have collected the shell at Aberdour, Largo Bay, Elie, etc.

APLYSIIDÆ.

Aplysia punctata (Cuv.).
> Firth of Forth (*M'B.*).

PLEUROPHYLLIDIIDÆ.

Pleurophyllidia lovéni (Bergh).
> Off Dunbar, 30 fathoms, in mud (*F. M. B.*).

V. NUDIBRANCHIATA—

HERMÆIDÆ.

Hermæa bifida (Mont.).
Black Rocks, Leith (*Landsb.*).

EOLIDIDÆ.

Eolis papillosa (Linn.).
Firth of Forth (*M'B., and Ed. Mus.*); not uncommon, Firth of Forth (*F. and H.*).
We have found this species frequently at Elie between tide marks.

E. coronata (Forb.).
On *Coryne decipiens*, North Queensferry (*M'B.*).

E. drummondi (Thomp.).
On *Tubularia indivisa*, North Queensferry (*M'B.*); Firth of Forth, 30 fathoms, and off Bass Rock, 24 fathoms (*Metz. and Mey.*).

E. landsburgi (Ald. and Han.).
On *Eudendrium rameum*, North Queensferry (*M'B.*).

E. nana (Ald. and Han.).
On *Hydractinia echinata*, Morrison's Haven (*T. S. W.*).

E. angulata (Ald. and Han.).
Off the Bass Rock, 24 fathoms (*Metz. and Mey.*).

E. despecta (Johnst.).
We dredged this small species east of Inchkeith last summer.

DOTONIDÆ.

Doto coronata (Gmel.).
Off Inchkeith (*M'B.*); shallow water, Firth of Forth (*F. M. B.*).

DENDRONOTIDÆ.

Dendronotus arborescens (O. F. Müll.).
Firth of Forth (*M'B., and Dr Grant*).
We have dredged this species near Inchkeith.

TRITONIIDÆ.

Tritonia hombergi (Cuv.).
Firth of Forth (*Flem., and Ed. Mus.*); Firth of Forth, 30 fathoms (*Metz. and Mey.*).

T. plebeia (Johnst.).
Firth of Forth (*M'B.*).
We have found this species at Elie.

POLYCERIDÆ.

Ægirus punctilucens (D'Orb.).
Shallow water, Firth of Forth (*F. M. B.*).

Triopa claviger (Müller).
Shallow water, Firth of Forth (*F. M. B.*).

Polycera quadrilineata (Müller).
Firth of Forth (*M'B.*).

P. lessoni (D'Orb.).
We dredged this species on the oyster bank last summer in 5 fathoms.

Ancula cristata (Alder).
Off Seafield, March 1857 (*M'B.*); Anstruther (*H. D. S. G.*).
We have found this species frequently at Elie, Wardie, and Aberdour, and have also dredged it.

Idalia aspersa (Ald. and Han.).
Off the Bass Rock, 24 fathoms (*Metz. and Mey.*).

Goniodoris nodosa (Mont.).
Between tide marks (*F. M. B.*).
We found several specimens lately at low water mark, west of Aberdour.

DORIDIDÆ.

Doris tuberculata (Cuv.).
> Firth of Forth (*M'B., and Ed. Mus.*); between tide marks (*F. M. B.*).
> This species is common in the littoral zone. We have found it at Wardie, North Berwick, Elie, Aberdour, etc.

D. bilamellata (Linn.).
> Firth of Forth (*M'B.*); abundant at low water in the Firth of Forth (*F. and H.*). Common at Aberdour, etc.

D. repanda (Ald. and Han.).
> We have taken this species frequently between tide marks at Aberdour, Elie, and Wardie.

D. pilosa (Müll.).
> Not uncommon on the shore at Aberdour, under stones, at low water.
> We lately took a specimen of the pure black variety of this species (the *Doris nigra* of Fleming) under a stone, at low water mark, on Carcraig Rock, near Inchcolm.

VI. PULMONOBRANCHIATA—

CARYCHIIDÆ.

Melampus bidentatus (Mont.).
> Occasionally on the beach at North Berwick (*M.*).

M. myosotis (Drap.), var. *ringens.*
> Rare on the beach at North Berwick (*M.*).

CEPHALOPODA.

I. DECAPODA—

TEUTHIDÆ.

Ommatostrephes todarus (Delle Chiaje).
> Firth of Forth (*Forb., F. and H.*); Leith

shores, common (*M'B.*); Granton and Craigroyston, cast ashore alive (*M.*).

This species is rather common in the Firth. We have found it alive at Portobello, Kinghorn, and Craigroyston, and have frequently obtained it dead, cast ashore.

Ommatostrephes sagittatus (Lamk.).

Firth of Forth (*Ed. Mus.*).

Loligo vulgaris (Lamk.).

Firth of Forth (*Dr Grant, F. and H.*): off Seafield, 1854 (*M'B.*); Craigroyston (*M.*). We have found this species frequently cast ashore at Portobello, Kinghorn, and Granton.

L. media (Linn.).

Aberlady Bay, 1857 (*M'B.*).

Rossia macrosoma (Delle Chiaje).

We obtained an adult individual of this species alive, at low water, on the Silver Sands, Aberdour, a few years ago.

Sepiola rondeleti (Leach).

Probably this is the species referred to under the name of *Loligo sepiola* by Dr Fleming, in his " British Animals," as having been found in the Firth of Forth by Dr Grant and also by himself.

II. OCTOPODA—

OCTOPIDÆ.

Octopus vulgaris (Lamk.).

Firth of Forth (*Grant*); not unfrequent in the Firth (*Neill*).

Probably it is this species which is recorded from the Forth by Dr Coldstream as *Octopus octopodia*.

Eledone cirrosa (Lamk.).

Kirkcaldy Bay, 1855 (*M'B.*).

APPENDIX.

Since the preceding pages have been printed we have received additional records of species from Mr Balfour, of Cambridge, and Professor Cunningham, of Belfast; these, along with others obtained by ourselves during a few days dredging in the neighbourhood of Inchcolm, and some previously omitted, are now inserted.

ALCYONARIA.

Virgularia mirabilis (Linn.).
We found this abundant in 5 to 10 fathoms, between Aberdour and Inchcolm, on a bottom of stiff blue mud.

OPHIURIDEA.

Amphiura filiformis (Müll.).
We have dredged this species on several occasions lately near Inchcolm, 18 fathoms.

HOLOTHUROIDEA.

Psolus phantapus (Linn.).
Professor R. O. Cunningham informs us that he has frequently obtained this species, taken at the entrance to the Firth, from the Prestonpans fishermen.

POLYZOA.

Bugula purpurotincta (Norman).
We have dredged this in considerable quantity on the west and north-west sides of Inchcolm, in 10 to 15 fathoms.

CIRRIPEDIA.

Balanus hameri (Ascanius).
Professor Cunningham writes, "I once obtained a magnificent mass attached to a stick, which it covered for about 2 feet."

AMPHIPODA.

Byblis gaimardi (Kröyer).
St Abb's Head, 40 fathoms (*Metz.*).

Corophium longicorne (Latr.).
    Very abundant in the mud flat at Morrison's Haven (*Cunningham*).
Dexamine spinosa (Leach).
    Low water, Prestonpans (*Cunningham*).
Hyperia galba (Mont.).
    In the pouches of *Medusæ* (*Cunningham*).
Ligia oceanica (Linn.).
    Under stones at high water mark (*Cunningham*).
    We have found it at Wardie.

CUMACEA.

Leucon nassica (Kröyer).
    St Abb's Head, 40 fathoms (*Metz.*).
Iphinoë gracilis (Bate).
    Bass Rock, 24 fathoms (*Metz.*).

STOMAPODA.

Themisto longispinosa (H. Goods.).
    Rocks, $2\frac{1}{2}$ fathoms, off Broxmouth, not uncommon (*F. M. B.*).

DECAPODA.

Hippolyte varians (Leach).
    Rocks off Broxmouth, near Dunbar (*F. M. B.*).
H. cranchii (Leach).
    In same locality as *H. varians* (*F. M. B.*).
H. thompsoni? (Bell).
    Firth of Forth (*F. M. B.*).
Pagurus cuanensis (Thompson).
    In *Turritella*, Firth of Forth (*F. M. B.*).
Ebalia cranchii (Leach).
    Mr Balfour dredged specimens twice in about 25 fathoms in mud about $2\frac{1}{2}$ miles off Dunbar.

TUNICATA.

Pelonaia corrugata (Forb. and Goods.).
    Beach at Portseaton after a gale, one specimen (*Cunningham*); not very uncommon (*F. M. B.*).

END.

www.ingramcontent.com/pod-product-compliance
Lightning Source LLC
Chambersburg PA
CBHW030407170426
43202CB00010B/1520